APPENDICES
TO ACCOMPANY

SECOND EDITION

FUNDAMENTALS OF ENGINEERING THERMODYNAMICS

MICHAEL J. MORAN
The Ohio State University

HOWARD N. SHAPIRO
Iowa State University of Science and Technology

JOHN WILEY & SONS, INC.
New York • Chichester • Brisbane • Toronto • Singapore

ISBN 0-471-56896-1

Printed and bound by Malloy Lithographing, Inc..

10 9 8 7 6 5 4 3 2 1

INDEX TO TABLES IN SI UNITS

TABLE A-1 Atomic or Molecular Weights and Critical Properties of Some Common Elements and Compounds

Substance	Chemical formula	M	T_c, K	p_c, bars	$Z_c = \dfrac{p_c v_c}{R T_c}$
Acetylene	C_2H_2	26.04	309	62.8	0.274
Air (equivalent)	—	28.97	133	37.7	0.284
Ammonia	NH_3	17.04	406	112.8	0.242
Argon	Ar	39.94	151	48.6	0.290
Benzene	C_6H_6	78.11	563	49.3	0.274
Butane	C_4H_{10}	58.12	425	38.0	0.274
Carbon	C	12.01	—	—	—
Carbon dioxide	CO_2	44.01	304	73.9	0.276
Carbon monoxide	CO	28.01	133	35.0	0.294
Copper	Cu	63.54	—	—	—
Ethane	C_2H_6	30.07	305	48.8	0.285
Ethyl alcohol	C_2H_5OH	46.07	516	63.8	0.249
Ethylene	C_2H_4	28.05	283	51.2	0.270
Helium	He	4.003	5.2	2.3	0.300
Hydrogen	H_2	2.018	33.2	13.0	0.304
Methane	CH_4	16.04	191	46.4	0.290
Methyl alcohol	CH_3OH	32.05	513	79.5	0.220
Nitrogen	N_2	28.01	126	33.9	0.291
Octane	C_8H_{18}	114.22	569	24.9	0.258
Oxygen	O_2	32.00	154	50.5	0.290
Propane	C_3H_8	44.09	370	42.7	0.276
Propylene	C_3H_6	42.08	365	46.2	0.276
Refrigerant 12	CCl_2F_2	120.92	385	41.2	0.278
Refrigerant 134a	CF_3CH_2F	102.03	374	40.7	0.260
Sulfur dioxide	SO_2	64.06	431	78.7	0.268
Water	H_2O	18.02	647.3	220.9	0.233

Sources: Adapted from *International Critical Tables* and L. C. Nelson and E. F. Obert, Generalized Compressibility Charts, *Chem. Eng., 61:* 203 (1954).

TABLE A-2 Properties of Saturated Water (Liquid–Vapor): Temperature Table

Temp. °C	Press. bars	Specific Volume m³/kg		Internal Energy kJ/kg		Enthalpy kJ/kg			Entropy kJ/kg · K		Temp. °C
		Sat. Liquid $v_f \times 10^3$	Sat. Vapor v_g	Sat. Liquid u_f	Sat. Vapor u_g	Sat. Liquid h_f	Evap. h_{fg}	Sat. Vapor h_g	Sat. Liquid s_f	Sat. Vapor s_g	
.01	0.00611	1.0002	206.136	0.00	2375.3	0.01	2501.3	2501.4	0.0000	9.1562	.01
4	0.00813	1.0001	157.232	16.77	2380.9	16.78	2491.9	2508.7	0.0610	9.0514	4
5	0.00872	1.0001	147.120	20.97	2382.3	20.98	2489.6	2510.6	0.0761	9.0257	5
6	0.00935	1.0001	137.734	25.19	2383.6	25.20	2487.2	2512.4	0.0912	9.0003	6
8	0.01072	1.0002	120.917	33.59	2386.4	33.60	2482.5	2516.1	0.1212	8.9501	8
10	0.01228	1.0004	106.379	42.00	2389.2	42.01	2477.7	2519.8	0.1510	8.9008	10
11	0.01312	1.0004	99.857	46.20	2390.5	46.20	2475.4	2521.6	0.1658	8.8765	11
12	0.01402	1.0005	93.784	50.41	2391.9	50.41	2473.0	2523.4	0.1806	8.8524	12
13	0.01497	1.0007	88.124	54.60	2393.3	54.60	2470.7	2525.3	0.1953	8.8285	13
14	0.01598	1.0008	82.848	58.79	2394.7	58.80	2468.3	2527.1	0.2099	8.8048	14
15	0.01705	1.0009	77.926	62.99	2396.1	62.99	2465.9	2528.9	0.2245	8.7814	15
16	0.01818	1.0011	73.333	67.18	2397.4	67.19	2463.6	2530.8	0.2390	8.7582	16
17	0.01938	1.0012	69.044	71.38	2398.8	71.38	2461.2	2532.6	0.2535	8.7351	17
18	0.02064	1.0014	65.038	75.57	2400.2	75.58	2458.8	2534.4	0.2679	8.7123	18
19	0.02198	1.0016	61.293	79.76	2401.6	79.77	2456.5	2536.2	0.2823	8.6897	19
20	0.02339	1.0018	57.791	83.95	2402.9	83.96	2454.1	2538.1	0.2966	8.6672	20
21	0.02487	1.0020	54.514	88.14	2404.3	88.14	2451.8	2539.9	0.3109	8.6450	21
22	0.02645	1.0022	51.447	92.32	2405.7	92.33	2449.4	2541.7	0.3251	8.6229	22
23	0.02810	1.0024	48.574	96.51	2407.0	96.52	2447.0	2543.5	0.3393	8.6011	23
24	0.02985	1.0027	45.883	100.70	2408.4	100.70	2444.7	2545.4	0.3534	8.5794	24
25	0.03169	1.0029	43.360	104.88	2409.8	104.89	2442.3	2547.2	0.3674	8.5580	25
26	0.03363	1.0032	40.994	109.06	2411.1	109.07	2439.9	2549.0	0.3814	8.5367	26
27	0.03567	1.0035	38.774	113.25	2412.5	113.25	2437.6	2550.8	0.3954	8.5156	27
28	0.03782	1.0037	36.690	117.42	2413.9	117.43	2435.2	2552.6	0.4093	8.4946	28
29	0.04008	1.0040	34.733	121.60	2415.2	121.61	2432.8	2554.5	0.4231	8.4739	29
30	0.04246	1.0043	32.894	125.78	2416.6	125.79	2430.5	2556.3	0.4369	8.4533	30
31	0.04496	1.0046	31.165	129.96	2418.0	129.97	2428.1	2558.1	0.4507	8.4329	31
32	0.04759	1.0050	29.540	134.14	2419.3	134.15	2425.7	2559.9	0.4644	8.4127	32
33	0.05034	1.0053	28.011	138.32	2420.7	138.33	2423.4	2561.7	0.4781	8.3927	33
34	0.05324	1.0056	26.571	142.50	2422.0	142.50	2421.0	2563.5	0.4917	8.3728	34
35	0.05628	1.0060	25.216	146.67	2423.4	146.68	2418.6	2565.3	0.5053	8.3531	35
36	0.05947	1.0063	23.940	150.85	2424.7	150.86	2416.2	2567.1	0.5188	8.3336	36
38	0.06632	1.0071	21.602	159.20	2427.4	159.21	2411.5	2570.7	0.5458	8.2950	38
40	0.07384	1.0078	19.523	167.56	2430.1	167.57	2406.7	2574.3	0.5725	8.2570	40
45	0.09593	1.0099	15.258	188.44	2436.8	188.45	2394.8	2583.2	0.6387	8.1648	45

TABLE A-2 (*Continued*)

Temp. °C	Press. bars	Specific Volume m³/kg		Internal Energy kJ/kg		Enthalpy kJ/kg			Entropy kJ/kg · K		Temp. °C
		Sat. Liquid $v_f \times 10^3$	Sat. Vapor v_g	Sat. Liquid u_f	Sat. Vapor u_g	Sat. Liquid h_f	Evap. h_{fg}	Sat. Vapor h_g	Sat. Liquid s_f	Sat. Vapor s_g	
50	.1235	1.0121	12.032	209.32	2443.5	209.33	2382.7	2592.1	.7038	8.0763	50
55	.1576	1.0146	9.568	230.21	2450.1	230.23	2370.7	2600.9	.7679	7.9913	55
60	.1994	1.0172	7.671	251.11	2456.6	251.13	2358.5	2609.6	.8312	7.9096	60
65	.2503	1.0199	6.197	272.02	2463.1	272.06	2346.2	2618.3	.8935	7.8310	65
70	.3119	1.0228	5.042	292.95	2469.6	292.98	2333.8	2626.8	.9549	7.7553	70
75	.3858	1.0259	4.131	313.90	2475.9	313.93	2321.4	2635.3	1.0155	7.6824	75
80	.4739	1.0291	3.407	334.86	2482.2	334.91	2308.8	2643.7	1.0753	7.6122	80
85	.5783	1.0325	2.828	355.84	2488.4	355.90	2296.0	2651.9	1.1343	7.5445	85
90	.7014	1.0360	2.361	376.85	2494.5	376.92	2283.2	2660.1	1.1925	7.4791	90
95	.8455	1.0397	1.982	397.88	2500.6	397.96	2270.2	2668.1	1.2500	7.4159	95
100	1.014	1.0435	1.673	418.94	2506.5	419.04	2257.0	2676.1	1.3069	7.3549	100
110	1.433	1.0516	1.210	461.14	2518.1	461.30	2230.2	2691.5	1.4185	7.2387	110
120	1.985	1.0603	0.8919	503.50	2529.3	503.71	2202.6	2706.3	1.5276	7.1296	120
130	2.701	1.0697	0.6685	546.02	2539.9	546.31	2174.2	2720.5	1.6344	7.0269	130
140	3.613	1.0797	0.5089	588.74	2550.0	589.13	2144.7	2733.9	1.7391	6.9299	140
150	4.758	1.0905	0.3928	631.68	2559.5	632.20	2114.3	2746.5	1.8418	6.8379	150
160	6.178	1.1020	0.3071	674.86	2568.4	675.55	2082.6	2758.1	1.9427	6.7502	160
170	7.917	1.1143	0.2428	718.33	2576.5	719.21	2049.5	2768.7	2.0419	6.6663	170
180	10.02	1.1274	0.1941	762.09	2583.7	763.22	2015.0	2778.2	2.1396	6.5857	180
190	12.54	1.1414	0.1565	806.19	2590.0	807.62	1978.8	2786.4	2.2359	6.5079	190
200	15.54	1.1565	0.1274	850.65	2595.3	852.45	1940.7	2793.2	2.3309	6.4323	200
210	19.06	1.1726	0.1044	895.53	2599.5	897.76	1900.7	2798.5	2.4248	6.3585	210
220	23.18	1.1900	0.08619	940.87	2602.4	943.62	1858.5	2802.1	2.5178	6.2861	220
230	27.95	1.2088	0.07158	986.74	2603.9	990.12	1813.8	2804.0	2.6099	6.2146	230
240	33.44	1.2291	0.05976	1033.2	2604.0	1037.3	1766.5	2803.8	2.7015	6.1437	240
250	39.73	1.2512	0.05013	1080.4	2602.4	1085.4	1716.2	2801.5	2.7927	6.0730	250
260	46.88	1.2755	0.04221	1128.4	2599.0	1134.4	1662.5	2796.6	2.8838	6.0019	260
270	54.99	1.3023	0.03564	1177.4	2593.7	1184.5	1605.2	2789.7	2.9751	5.9301	270
280	64.12	1.3321	0.03017	1227.5	2586.1	1236.0	1543.6	2779.6	3.0668	5.8571	280
290	74.36	1.3656	0.02557	1278.9	2576.0	1289.1	1477.1	2766.2	3.1594	5.7821	290
300	85.81	1.4036	0.02167	1332.0	2563.0	1344.0	1404.9	2749.0	3.2534	5.7045	300
320	112.7	1.4988	0.01549	1444.6	2525.5	1461.5	1238.6	2700.1	3.4480	5.5362	320
340	145.9	1.6379	0.01080	1570.3	2464.6	1594.2	1027.9	2622.0	3.6594	5.3357	340
360	186.5	1.8925	0.006945	1725.2	2351.5	1760.5	720.5	2481.0	3.9147	5.0526	360
374.14	220.9	3.155	0.003155	2029.6	2029.6	2099.3	0	2099.3	4.4298	4.4298	374.14

Source: Tables A-2 through A-5 are adapted from K. Wark, *Thermodynamics, 4th ed.,* McGraw-Hill, New York, 1983, as extracted from J. H. Keenan, F. G. Keyes, P. G. Hill, and J. G. Moore, *Steam Tables,* Wiley, New York, 1969.

TABLE A-3 Properties of Saturated Water (Liquid–Vapor): Pressure Table

Press. bars	Temp. °C	Specific Volume m³/kg		Internal Energy kJ/kg		Enthalpy kJ/kg			Entropy kJ/kg · K		Press. bars
		Sat. Liquid $v_f \times 10^3$	Sat. Vapor v_g	Sat. Liquid u_f	Sat. Vapor u_g	Sat. Liquid h_f	Evap. h_{fg}	Sat. Vapor h_g	Sat. Liquid s_f	Sat. Vapor s_g	
0.04	28.96	1.0040	34.800	121.45	2415.2	121.46	2432.9	2554.4	0.4226	8.4746	0.04
0.06	36.16	1.0064	23.739	151.53	2425.0	151.53	2415.9	2567.4	0.5210	8.3304	0.06
0.08	41.51	1.0084	18.103	173.87	2432.2	173.88	2403.1	2577.0	0.5926	8.2287	0.08
0.10	45.81	1.0102	14.674	191.82	2437.9	191.83	2392.8	2584.7	0.6493	8.1502	0.10
0.20	60.06	1.0172	7.649	251.38	2456.7	251.40	2358.3	2609.7	0.8320	7.9085	0.20
0.30	69.10	1.0223	5.229	289.20	2468.4	289.23	2336.1	2625.3	0.9439	7.7686	0.30
0.40	75.87	1.0265	3.993	317.53	2477.0	317.58	2319.2	2636.8	1.0259	7.6700	0.40
0.50	81.33	1.0300	3.240	340.44	2483.9	340.49	2305.4	2645.9	1.0910	7.5939	0.50
0.60	85.94	1.0331	2.732	359.79	2489.6	359.86	2293.6	2653.5	1.1453	7.5320	0.60
0.70	89.95	1.0360	2.365	376.63	2494.5	376.70	2283.3	2660.0	1.1919	7.4797	0.70
0.80	93.50	1.0380	2.087	391.58	2498.8	391.66	2274.1	2665.8	1.2329	7.4346	0.80
0.90	96.71	1.0410	1.869	405.06	2502.6	405.15	2265.7	2670.9	1.2695	7.3949	0.90
1.00	99.63	1.0432	1.694	417.36	2506.1	417.46	2258.0	2675.5	1.3026	7.3594	1.00
1.50	111.4	1.0528	1.159	466.94	2519.7	467.11	2226.5	2693.6	1.4336	7.2233	1.50
2.00	120.2	1.0605	0.8857	504.49	2529.5	504.70	2201.9	2706.7	1.5301	7.1271	2.00
2.50	127.4	1.0672	0.7187	535.10	2537.2	535.37	2181.5	2716.9	1.6072	7.0527	2.50
3.00	133.6	1.0732	0.6058	561.15	2543.6	561.47	2163.8	2725.3	1.6718	6.9919	3.00
3.50	138.9	1.0786	0.5243	583.95	2546.9	584.33	2148.1	2732.4	1.7275	6.9405	3.50
4.00	143.6	1.0836	0.4625	604.31	2553.6	604.74	2133.8	2738.6	1.7766	6.8959	4.00
4.50	147.9	1.0882	0.4140	622.25	2557.6	623.25	2120.7	2743.9	1.8207	6.8565	4.50
5.00	151.9	1.0926	0.3749	639.68	2561.2	640.23	2108.5	2748.7	1.8607	6.8212	5.00
6.00	158.9	1.1006	0.3157	669.90	2567.4	670.56	2086.3	2756.8	1.9312	6.7600	6.00
7.00	165.0	1.1080	0.2729	696.44	2572.5	697.22	2066.3	2763.5	1.9922	6.7080	7.00
8.00	170.4	1.1148	0.2404	720.22	2576.8	721.11	2048.0	2769.1	2.0462	6.6628	8.00
9.00	175.4	1.1212	0.2150	741.83	2580.5	742.83	2031.1	2773.9	2.0946	6.6226	9.00
10.0	179.9	1.1273	0.1944	761.68	2583.6	762.81	2015.3	2778.1	2.1387	6.5863	10.0
15.0	198.3	1.1539	0.1318	843.16	2594.5	844.84	1947.3	2792.2	2.3150	6.4448	15.0
20.0	212.4	1.1767	0.09963	906.44	2600.3	908.79	1890.7	2799.5	2.4474	6.3409	20.0
25.0	224.0	1.1973	0.07998	959.11	2603.1	962.11	1841.0	2803.1	2.5547	6.2575	25.0
30.0	233.9	1.2165	0.06668	1004.8	2604.1	1008.4	1795.7	2804.2	2.6457	6.1869	30.0
35.0	242.6	1.2347	0.05707	1045.4	2603.7	1049.8	1753.7	2803.4	2.7253	6.1253	35.0
40.0	250.4	1.2522	0.04978	1082.3	2602.3	1087.3	1714.1	2801.4	2.7964	6.0701	40.0
45.0	257.5	1.2692	0.04406	1116.2	2600.1	1121.9	1676.4	2798.3	2.8610	6.0199	45.0
50.0	264.0	1.2859	0.03944	1147.8	2597.1	1154.2	1640.1	2794.3	2.9202	5.9734	50.0
60.0	275.6	1.3187	0.03244	1205.4	2589.7	1213.4	1571.0	2784.3	3.0267	5.8892	60.0
70.0	285.9	1.3513	0.02737	1257.6	2580.5	1267.0	1505.1	2772.1	3.1211	5.8133	70.0
80.0	295.1	1.3842	0.02352	1305.6	2569.8	1316.6	1441.3	2758.0	3.2068	5.7432	80.0
90.0	303.4	1.4178	0.02048	1350.5	2557.8	1363.3	1378.9	2742.1	3.2858	5.6772	90.0
100.	311.1	1.4524	0.01803	1393.0	2544.4	1407.6	1317.1	2724.7	3.3596	5.6141	100.
110.	318.2	1.4886	0.01599	1433.7	2529.8	1450.1	1255.5	2705.6	3.4295	5.5527	110.

TABLE A-3 (*Continued*)

Press. bars	Temp. °C	Specific Volume m³/kg		Internal Energy kJ/kg		Enthalpy kJ/kg			Entropy kJ/kg · K		Press. bars
		Sat. Liquid $v_f \times 10^3$	Sat. Vapor v_g	Sat. Liquid u_f	Sat. Vapor u_g	Sat. Liquid h_f	Evap. h_{fg}	Sat. Vapor h_g	Sat. Liquid s_f	Sat. Vapor s_g	
120.	324.8	1.5267	0.01426	1473.0	2513.7	1491.3	1193.6	2684.9	3.4962	5.4924	120.
130.	330.9	1.5671	0.01278	1511.1	2496.1	1531.5	1130.7	2662.2	3.5606	5.4323	130.
140.	336.8	1.6107	0.01149	1548.6	2476.8	1571.1	1066.5	2637.6	3.6232	5.3717	140.
150.	342.2	1.6581	0.01034	1585.6	2455.5	1610.5	1000.0	2610.5	3.6848	5.3098	150.
160.	347.4	1.7107	0.009306	1622.7	2431.7	1650.1	930.6	2580.6	3.7461	5.2455	160.
170.	352.4	1.7702	0.008364	1660.2	2405.0	1690.3	856.9	2547.2	3.8079	5.1777	170.
180.	357.1	1.8397	0.007489	1698.9	2374.3	1732.0	777.1	2509.1	3.8715	5.1044	180.
190.	361.5	1.9243	0.006657	1739.9	2338.1	1776.5	688.0	2464.5	3.9388	5.0228	190.
200.	365.8	2.036	0.005834	1785.6	2293.0	1826.3	583.4	2409.7	4.0139	4.9269	200.
220.9	374.1	3.155	0.003155	2029.6	2029.6	2099.3	0	2099.3	4.4298	4.4298	220.9

TABLE A-4 Properties of Superheated Water Vapor

T °C	v m³/kg	u kJ/kg	h kJ/kg	s kJ/kg · K	v m³/kg	u kJ/kg	h kJ/kg	s kJ/kg · K
	$p = 0.06$ bar $= 0.006$ MPa ($T_{sat} = 36.16$°C)				$p = 0.35$ bar $= 0.035$ MPa ($T_{sat} = 72.69$°C)			
Sat.	23.739	2425.0	2567.4	8.3304	4.526	2473.0	2631.4	7.7158
80	27.132	2487.3	2650.1	8.5804	4.625	2483.7	2645.6	7.7564
120	30.219	2544.7	2726.0	8.7840	5.163	2542.4	2723.1	7.9644
160	33.302	2602.7	2802.5	8.9693	5.696	2601.2	2800.6	8.1519
200	36.383	2661.4	2879.7	9.1398	6.228	2660.4	2878.4	8.3237
240	39.462	2721.0	2957.8	9.2982	6.758	2720.3	2956.8	8.4828
280	42.540	2781.5	3036.8	9.4464	7.287	2780.9	3036.0	8.6314
320	45.618	2843.0	3116.7	9.5859	7.815	2842.5	3116.1	8.7712
360	48.696	2905.5	3197.7	9.7180	8.344	2905.1	3197.1	8.9034
400	51.774	2969.0	3279.6	9.8435	8.872	2968.6	3279.2	9.0291
440	54.851	3033.5	3362.6	9.9633	9.400	3033.2	3362.2	9.1490
500	59.467	3132.3	3489.1	10.1336	10.192	3132.1	3488.8	9.3194

T °C	v m³/kg	u kJ/kg	h kJ/kg	s kJ/kg · K	v m³/kg	u kJ/kg	h kJ/kg	s kJ/kg · K
	$p = 0.70$ bar $= 0.07$ MPa ($T_{sat} = 89.95$°C)				$p = 1.0$ bar $= 0.10$ MPa ($T_{sat} = 99.63$°C)			
Sat.	2.365	2494.5	2660.0	7.4797	1.694	2506.1	2675.5	7.3594
100	2.434	2509.7	2680.0	7.5341	1.696	2506.7	2676.2	7.3614
120	2.571	2539.7	2719.6	7.6375	1.793	2537.3	2716.6	7.4668
160	2.841	2599.4	2798.2	7.8279	1.984	2597.8	2796.2	7.6597
200	3.108	2659.1	2876.7	8.0012	2.172	2658.1	2875.3	7.8343
240	3.374	2719.3	2955.5	8.1611	2.359	2718.5	2954.5	7.9949
280	3.640	2780.2	3035.0	8.3162	2.546	2779.6	3034.2	8.1445
320	3.905	2842.0	3115.3	8.4504	2.732	2841.5	3114.6	8.2849
360	4.170	2904.6	3196.5	8.5828	2.917	2904.2	3195.9	8.4175
400	4.434	2968.2	3278.6	8.7086	3.103	2967.9	3278.2	8.5435
440	4.698	3032.9	3361.8	8.8286	3.288	3032.6	3361.4	8.6636
500	5.095	3131.8	3488.5	8.9991	3.565	3131.6	3488.1	8.8342

T °C	v m³/kg	u kJ/kg	h kJ/kg	s kJ/kg · K	v m³/kg	u kJ/kg	h kJ/kg	s kJ/kg · K
	$p = 1.5$ bars $= 0.15$ MPa ($T_{sat} = 111.37$°C)				$p = 3.0$ bars $= 0.30$ MPa ($T_{sat} = 133.55$°C)			
Sat.	1.159	2519.7	2693.6	7.2233	0.606	2543.6	2725.3	6.9919
120	1.188	2533.3	2711.4	7.2693				
160	1.317	2595.2	2792.8	7.4665	0.651	2587.1	2782.3	7.1276
200	1.444	2656.2	2872.9	7.6433	0.716	2650.7	2865.5	7.3115
240	1.570	2717.2	2952.7	7.8052	0.781	2713.1	2947.3	7.4774
280	1.695	2778.6	3032.8	7.9555	0.844	2775.4	3028.6	7.6299
320	1.819	2840.6	3113.5	8.0964	0.907	2838.1	3110.1	7.7722
360	1.943	2903.5	3195.0	8.2293	0.969	2901.4	3192.2	7.9061
400	2.067	2967.3	3277.4	8.3555	1.032	2965.6	3275.0	8.0330
440	2.191	3032.1	3360.7	8.4757	1.094	3030.6	3358.7	8.1538
500	2.376	3131.2	3487.6	8.6466	1.187	3130.0	3486.0	8.3251
600	2.685	3301.7	3704.3	8.9101	1.341	3300.8	3703.2	8.5892

TABLE A-4 (*Continued*)

T °C	v m³/kg	u kJ/kg	h kJ/kg	s kJ/kg · K	v m³/kg	u kJ/kg	h kJ/kg	s kJ/kg · K
	p = 5.0 bars = 0.50 MPa (*T*ₛₐₜ = 151.86°C)				*p* = 7.0 bars = 0.70 MPa (*T*ₛₐₜ = 164.97°C)			
Sat.	0.3749	2561.2	2748.7	6.8213	0.2729	2572.5	2763.5	6.7080
180	0.4045	2609.7	2812.0	6.9656	0.2847	2599.8	2799.1	6.7880
200	0.4249	2642.9	2855.4	7.0592	0.2999	2634.8	2844.8	6.8865
240	0.4646	2707.6	2939.9	7.2307	0.3292	2701.8	2932.2	7.0641
280	0.5034	2771.2	3022.9	7.3865	0.3574	2766.9	3017.1	7.2233
320	0.5416	2834.7	3105.6	7.5308	0.3852	2831.3	3100.9	7.3697
360	0.5796	2898.7	3188.4	7.6660	0.4126	2895.8	3184.7	7.5063
400	0.6173	2963.2	3271.9	7.7938	0.4397	2960.9	3268.7	7.6350
440	0.6548	3028.6	3356.0	7.9152	0.4667	3026.6	3353.3	7.7571
500	0.7109	3128.4	3483.9	8.0873	0.5070	3126.8	3481.7	7.9299
600	0.8041	3299.6	3701.7	8.3522	0.5738	3298.5	3700.2	8.1956
700	0.8969	3477.5	3925.9	8.5952	0.6403	3476.6	3924.8	8.4391
	p = 10.0 bars = 1.0 MPa (*T*ₛₐₜ = 179.91°C)				*p* = 15.0 bars = 1.5 MPa (*T*ₛₐₜ = 198.32°C)			
Sat.	0.1944	2583.6	2778.1	6.5865	0.1318	2594.5	2792.2	6.4448
200	0.2060	2621.9	2827.9	6.6940	0.1325	2598.1	2796.8	6.4546
240	0.2275	2692.9	2920.4	6.8817	0.1483	2676.9	2899.3	6.6628
280	0.2480	2760.2	3008.2	7.0465	0.1627	2748.6	2992.7	6.8381
320	0.2678	2826.1	3093.9	7.1962	0.1765	2817.1	3081.9	6.9938
360	0.2873	2891.6	3178.9	7.3349	0.1899	2884.4	3169.2	7.1363
400	0.3066	2957.3	3263.9	7.4651	0.2030	2951.3	3255.8	7.2690
440	0.3257	3023.6	3349.3	7.5883	0.2160	3018.5	3342.5	7.3940
500	0.3541	3124.4	3478.5	7.7622	0.2352	3120.3	3473.1	7.5698
540	0.3729	3192.6	3565.6	7.8720	0.2478	3189.1	3560.9	7.6805
600	0.4011	3296.8	3697.9	8.0290	0.2668	3293.9	3694.0	7.8385
640	0.4198	3367.4	3787.2	8.1290	0.2793	3364.8	3783.8	7.9391
	p = 20.0 bars = 2.0 MPa (*T*ₛₐₜ = 212.42°C)				*p* = 30.0 bars = 3.0 MPa (*T*ₛₐₜ = 233.90°C)			
Sat.	0.0996	2600.3	2799.5	6.3409	0.0667	2604.1	2804.2	6.1869
240	0.1085	2659.6	2876.5	6.4952	0.0682	2619.7	2824.3	6.2265
280	0.1200	2736.4	2976.4	6.6828	0.0771	2709.9	2941.3	6.4462
320	0.1308	2807.9	3069.5	6.8452	0.0850	2788.4	3043.4	6.6245
360	0.1411	2877.0	3159.3	6.9917	0.0923	2861.7	3138.7	6.7801
400	0.1512	2945.2	3247.6	7.1271	0.0994	2932.8	3230.9	6.9212
440	0.1611	3013.4	3335.5	7.2540	0.1062	3002.9	3321.5	7.0520
500	0.1757	3116.2	3467.6	7.4317	0.1162	3108.0	3456.5	7.2338
540	0.1853	3185.6	3556.1	7.5434	0.1227	3178.4	3546.6	7.3474
600	0.1996	3290.9	3690.1	7.7024	0.1324	3285.0	3682.3	7.5085
640	0.2091	3362.2	3780.4	7.8035	0.1388	3357.0	3773.5	7.6106
700	0.2232	3470.9	3917.4	7.9487	0.1484	3466.5	3911.7	7.7571

TABLE A-4 (*Continued*)

T °C	v m³/kg	u kJ/kg	h kJ/kg	s kJ/kg · K	v m³/kg	u kJ/kg	h kJ/kg	s kJ/kg · K
	p = 40 bars = 4.0 MPa (T_{sat} = 250.4°C)				*p* = 60 bars = 6.0 MPa (T_{sat} = 275.64°C)			
Sat.	0.04978	2602.3	2801.4	6.0701	0.03244	2589.7	2784.3	5.8892
280	0.05546	2680.0	2901.8	6.2568	0.03317	2605.2	2804.2	5.9252
320	0.06199	2767.4	3015.4	6.4553	0.03876	2720.0	2952.6	6.1846
360	0.06788	2845.7	3117.2	6.6215	0.04331	2811.2	3071.1	6.3782
400	0.07341	2919.9	3213.6	6.7690	0.04739	2892.9	3177.2	6.5408
440	0.07872	2992.2	3307.1	6.9041	0.05122	2970.0	3277.3	6.6853
500	0.08643	3099.5	3445.3	7.0901	0.05665	3082.2	3422.2	6.8803
540	0.09145	3171.1	3536.9	7.2056	0.06015	3156.1	3517.0	6.9999
600	0.09885	3279.1	3674.4	7.3688	0.06525	3266.9	3658.4	7.1677
640	0.1037	3351.8	3766.6	7.4720	0.06859	3341.0	3752.6	7.2731
700	0.1110	3462.1	3905.9	7.6198	0.07352	3453.1	3894.1	7.4234
740	0.1157	3536.6	3999.6	7.7141	0.07677	3528.3	3989.2	7.5190
	p = 80 bars = 8.0 MPa (T_{sat} = 295.06°C)				*p* = 100 bars = 10.0 MPa (T_{sat} = 311.06°C)			
Sat.	0.02352	2569.8	2758.0	5.7432	0.01803	2544.4	2724.7	5.6141
320	0.02682	2662.7	2877.2	5.9489	0.01925	2588.8	2781.3	5.7103
360	0.03089	2772.7	3019.8	6.1819	0.02331	2729.1	2962.1	6.0060
400	0.03432	2863.8	3138.3	6.3634	0.02641	2832.4	3096.5	6.2120
440	0.03742	2946.7	3246.1	6.5190	0.02911	2922.1	3213.2	6.3805
480	0.04034	3025.7	3348.4	6.6586	0.03160	3005.4	3321.4	6.5282
520	0.04313	3102.7	3447.7	6.7871	0.03394	3085.6	3425.1	6.6622
560	0.04582	3178.7	3545.3	6.9072	0.03619	3164.1	3526.0	6.7864
600	0.04845	3254.4	3642.0	7.0206	0.03837	3241.7	3625.3	6.9029
640	0.05102	3330.1	3738.3	7.1283	0.04048	3318.9	3723.7	7.0131
700	0.05481	3443.9	3882.4	7.2812	0.04358	3434.7	3870.5	7.1687
740	0.05729	3520.4	3978.7	7.3782	0.04560	3512.1	3968.1	7.2670
	p = 120 bars = 12.0 MPa (T_{sat} = 324.75°C)				*p* = 140 bars = 14.0 MPa (T_{sat} = 336.75°C)			
Sat.	0.01426	2513.7	2684.9	5.4924	0.01149	2476.8	2637.6	5.3717
360	0.01811	2678.4	2895.7	5.8361	0.01422	2617.4	2816.5	5.6602
400	0.02108	2798.3	3051.3	6.0747	0.01722	2760.9	3001.9	5.9448
440	0.02355	2896.1	3178.7	6.2586	0.01954	2868.6	3142.2	6.1474
480	0.02576	2984.4	3293.5	6.4154	0.02157	2962.5	3264.5	6.3143
520	0.02781	3068.0	3401.8	6.5555	0.02343	3049.8	3377.8	6.4610
560	0.02977	3149.0	3506.2	6.6840	0.02517	3133.6	3486.0	6.5941
600	0.03164	3228.7	3608.3	6.8037	0.02683	3215.4	3591.1	6.7172
640	0.03345	3307.5	3709.0	6.9164	0.02843	3296.0	3694.1	6.8326
700	0.03610	3425.2	3858.4	7.0749	0.03075	3415.7	3846.2	6.9939
740	0.03781	3503.7	3957.4	7.1746	0.03225	3495.2	3946.7	7.0952

TABLE A-4 (*Continued*)

T °C	v m³/kg	u kJ/kg	h kJ/kg	s kJ/kg · K	v m³/kg	u kJ/kg	h kJ/kg	s kJ/kg · K
	p = 160 bars = 16.0 MPa (T_{sat} = 347.44°C)				*p* = 180 bars = 18.0 MPa (T_{sat} = 357.06°C)			
Sat.	0.00931	2431.7	2580.6	5.2455	0.00749	2374.3	2509.1	5.1044
360	0.01105	2539.0	2715.8	5.4614	0.00809	2418.9	2564.5	5.1922
400	0.01426	2719.4	2947.6	5.8175	0.01190	2672.8	2887.0	5.6887
440	0.01652	2839.4	3103.7	6.0429	0.01414	2808.2	3062.8	5.9428
480	0.01842	2939.7	3234.4	6.2215	0.01596	2915.9	3203.2	6.1345
520	0.02013	3031.1	3353.3	6.3752	0.01757	3011.8	3378.0	6.2960
560	0.02172	3117.8	3465.4	6.5132	0.01904	3101.7	3444.4	6.4392
600	0.02323	3201.8	3573.5	6.6399	0.02042	3188.0	3555.6	6.5696
640	0.02467	3284.2	3678.9	6.7580	0.02174	3272.3	3663.6	6.6905
700	0.02674	3406.0	3833.9	6.9224	0.02362	3396.3	3821.5	6.8580
740	0.02808	3486.7	3935.9	7.0251	0.02483	3478.0	3925.0	6.9623

T °C	v m³/kg	u kJ/kg	h kJ/kg	s kJ/kg · K	v m³/kg	u kJ/kg	h kJ/kg	s kJ/kg · K
	p = 200 bars = 20.0 MPa (T_{sat} = 365.81°C)				*p* = 240 bars = 24.0 MPa			
Sat.	0.00583	2293.0	2409.7	4.9269				
400	0.00994	2619.3	2818.1	5.5540	0.00673	2477.8	2639.4	5.2393
440	0.01222	2774.9	3019.4	5.8450	0.00929	2700.6	2923.4	5.6506
480	0.01399	2891.2	3170.8	6.0518	0.01100	2838.3	3102.3	5.8950
520	0.01551	2992.0	3302.2	6.2218	0.01241	2950.5	3248.5	6.0842
560	0.01689	3085.2	3423.0	6.3705	0.01366	3051.1	3379.0	6.2448
600	0.01818	3174.0	3537.6	6.5048	0.01481	3145.2	3500.7	6.3875
640	0.01940	3260.2	3648.1	6.6286	0.01588	3235.5	3616.7	6.5174
700	0.02113	3386.4	3809.0	6.7993	0.01739	3366.4	3783.8	6.6947
740	0.02224	3469.3	3914.1	6.9052	0.01835	3451.7	3892.1	6.8038
800	0.02385	3592.7	4069.7	7.0544	0.01974	3578.0	4051.6	6.9567

T °C	v m³/kg	u kJ/kg	h kJ/kg	s kJ/kg · K	v m³/kg	u kJ/kg	h kJ/kg	s kJ/kg · K
	p = 280 bars = 28.0 MPa				*p* = 320 bars = 32.0 MPa			
400	0.00383	2223.5	2330.7	4.7494	0.00236	1980.4	2055.9	4.3239
440	0.00712	2613.2	2812.6	5.4494	0.00544	2509.0	2683.0	5.2327
480	0.00885	2780.8	3028.5	5.7446	0.00722	2718.1	2949.2	5.5968
520	0.01020	2906.8	3192.3	5.9566	0.00853	2860.7	3133.7	5.8357
560	0.01136	3015.7	3333.7	6.1307	0.00963	2979.0	3287.2	6.0246
600	0.01241	3115.6	3463.0	6.2823	0.01061	3085.3	3424.6	6.1858
640	0.01338	3210.3	3584.8	6.4187	0.01150	3184.5	3552.5	6.3290
700	0.01473	3346.1	3758.4	6.6029	0.01273	3325.4	3732.8	6.5203
740	0.01558	3433.9	3870.0	6.7153	0.01350	3415.9	3847.8	6.6361
800	0.01680	3563.1	4033.4	6.8720	0.01460	3548.0	4015.1	6.7966
900	0.01873	3774.3	4298.8	7.1084	0.01633	3762.7	4285.1	7.0372

TABLE A-5 Properties of Compressed Liquid Water

T °C	$v \times 10^3$ m³/kg	u kJ/kg	h kJ/kg	s kJ/kg·K	$v \times 10^3$ m³/kg	u kJ/kg	h kJ/kg	s kJ/kg·K
	p = 25 bars = 2.5 MPa (T_{sat} = 223.99°C)				p = 50 bars = 5.0 MPa (T_{sat} = 263.99°C)			
20	1.0006	83.80	86.30	.2961	.9995	83.65	88.65	.2956
40	1.0067	167.25	169.77	.5715	1.0056	166.95	171.97	.5705
80	1.0280	334.29	336.86	1.0737	1.0268	333.72	338.85	1.0720
100	1.0423	418.24	420.85	1.3050	1.0410	417.52	422.72	1.3030
140	1.0784	587.82	590.52	1.7369	1.0768	586.76	592.15	1.7343
180	1.1261	761.16	763.97	2.1375	1.1240	759.63	765.25	2.1341
200	1.1555	849.9	852.8	2.3294	1.1530	848.1	853.9	2.3255
220	1.1898	940.7	943.7	2.5174	1.1866	938.4	944.4	2.5128
Sat.	1.1973	959.1	962.1	2.5546	1.2859	1147.8	1154.2	2.9202
	p = 75 bars = 7.5 MPa (T_{sat} = 290.59°C)				p = 100 bars = 10.0 MPa (T_{sat} = 311.06°C)			
20	.9984	83.50	90.99	.2950	.9972	83.36	93.33	.2945
40	1.0045	166.64	174.18	.5696	1.0034	166.35	176.38	.5686
80	1.0256	333.15	340.84	1.0704	1.0245	332.59	342.83	1.0688
100	1.0397	416.81	424.62	1.3011	1.0385	416.12	426.50	1.2992
140	1.0752	585.72	593.78	1.7317	1.0737	584.68	595.42	1.7292
180	1.1219	758.13	766.55	2.1308	1.1199	756.65	767.84	2.1275
220	1.1835	936.2	945.1	2.5083	1.1805	934.1	945.9	2.5039
260	1.2696	1124.4	1134.0	2.8763	1.2645	1121.1	1133.7	2.8699
Sat.	1.3677	1282.0	1292.2	3.1649	1.4524	1393.0	1407.6	3.3596
	p = 150 bars = 15.0 MPa (T_{sat} = 342.24°C)				p = 200 bars = 20.0 MPa (T_{sat} = 365.81°C)			
20	.9950	83.06	97.99	.2934	.9928	82.77	102.62	.2923
40	1.0013	165.76	180.78	.5666	.9992	165.17	185.16	.5646
80	1.0222	331.48	346.81	1.0656	1.0199	330.40	350.80	1.0624
100	1.0361	414.74	430.28	1.2955	1.0337	413.39	434.06	1.2917
140	1.0707	582.66	598.72	1.7242	1.0678	580.69	602.04	1.7193
180	1.1159	753.76	770.50	2.1210	1.1120	750.95	773.20	2.1147
220	1.1748	929.9	947.5	2.4953	1.1693	925.9	949.3	2.4870
260	1.2550	1114.6	1133.4	2.8576	1.2462	1108.6	1133.5	2.8459
300	1.3770	1316.6	1337.3	3.2260	1.3596	1306.1	1333.3	3.2071
Sat.	1.6581	1585.6	1610.5	3.6848	2.036	1785.6	1826.3	4.0139
	p = 250 bars = 25 MPa				p = 300 bars = 30 MPa			
20	.9907	82.47	107.24	.2911	.9886	82.17	111.84	.2899
40	.9971	164.60	189.52	.5626	.9951	164.04	193.89	.5607
100	1.0313	412.08	437.85	1.2881	1.0290	410.78	441.66	1.2844
200	1.1344	834.5	862.8	2.2961	1.1302	831.4	865.3	2.2893
300	1.3442	1296.6	1330.2	3.1900	1.3304	1287.9	1327.8	3.1741

TABLE A-6 Properties of Saturated Water (Solid–Vapor): Temperature Table

Temp. °C	Pressure kPa	Specific Volume m³/kg		Internal Energy kJ/kg			Enthalpy kJ/kg			Entropy kJ/kg·K		
		Sat. Solid $v_i \times 10^3$	Sat. Vapor v_g	Sat. Solid u_i	Subl. u_{ig}	Sat. Vapor u_g	Sat. Solid h_i	Subl. h_{ig}	Sat. Vapor h_g	Sat. Solid s_i	Subl. s_{ig}	Sat. Vapor s_g
.01	.6113	1.0908	206.1	−333.40	2708.7	2375.3	−333.40	2834.8	2501.4	−1.221	10.378	9.156
0	.6108	1.0908	206.3	−333.43	2708.8	2375.3	−333.43	2834.8	2501.3	−1.221	10.378	9.157
−2	.5176	1.0904	241.7	−337.62	2710.2	2372.6	−337.62	2835.3	2497.7	−1.237	10.456	9.219
−4	.4375	1.0901	283.8	−341.78	2711.6	2369.8	−341.78	2835.7	2494.0	−1.253	10.536	9.283
−6	.3689	1.0898	334.2	−345.91	2712.9	2367.0	−345.91	2836.2	2490.3	−1.268	10.616	9.348
−8	.3102	1.0894	394.4	−350.02	2714.2	2364.2	−350.02	2836.6	2486.6	−1.284	10.698	9.414
−10	.2602	1.0891	466.7	−354.09	2715.5	2361.4	−354.09	2837.0	2482.9	−1.299	10.781	9.481
−12	.2176	1.0888	553.7	−358.14	2716.8	2358.7	−358.14	2837.3	2479.2	−1.315	10.865	9.550
−14	.1815	1.0884	658.8	−362.15	2718.0	2355.9	−362.15	2837.6	2475.5	−1.331	10.950	9.619
−16	.1510	1.0881	786.0	−366.14	2719.2	2353.1	−366.14	2837.9	2471.8	−1.346	11.036	9.690
−18	.1252	1.0878	940.5	−370.10	2720.4	2350.3	−370.10	2838.2	2468.1	−1.362	11.123	9.762
−20	.1035	1.0874	1128.6	−374.03	2721.6	2347.5	−374.03	2838.4	2464.3	−1.377	11.212	9.835
−22	.0853	1.0871	1358.4	−377.93	2722.7	2344.7	−377.93	2838.6	2460.6	−1.393	11.302	9.909
−24	.0701	1.0868	1640.1	−381.80	2723.7	2342.0	−381.80	2838.7	2456.9	−1.408	11.394	9.985
−26	.0574	1.0864	1986.4	−385.64	2724.8	2339.2	−385.64	2838.9	2453.2	−1.424	11.486	10.062
−28	.0469	1.0861	2413.7	−389.45	2725.8	2336.4	−389.45	2839.0	2449.5	−1.439	11.580	10.141
−30	.0381	1.0858	2943.	−393.23	2726.8	2333.6	−393.23	2839.0	2445.8	−1.455	11.676	10.221
−32	.0309	1.0854	3600.	−396.98	2727.8	2330.8	−396.98	2839.1	2442.1	−1.471	11.773	10.303
−34	.0250	1.0851	4419.	−400.71	2728.7	2328.0	−400.71	2839.1	2438.4	−1.486	11.872	10.386
−36	.0201	1.0848	5444.	−404.40	2729.6	2325.2	−404.40	2839.1	2434.7	−1.501	11.972	10.470
−38	.0161	1.0844	6731.	−408.06	2730.5	2322.4	−408.06	2839.0	2430.9	−1.517	12.073	10.556
−40	.0129	1.0841	8354.	−411.70	2731.3	2319.6	−411.70	2838.9	2427.2	−1.532	12.176	10.644

Source: J. H. Keenan, F. G. Keyes, P. G. Hill, and J. G. Moore, *Steam Tables*, Wiley, New York, 1978.

TABLE A-7 Properties of Saturated Refrigerant 12 (Liquid–Vapor): Temperature Table

Temp. °C	Press. bars	Specific Volume m³/kg		Internal Energy kJ/kg		Enthalpy kJ/kg			Entropy kJ/kg · K		Temp. °C
		Sat. Liquid $v_f \times 10^3$	Sat. Vapor v_g	Sat. Liquid u_f	Sat. Vapor u_g	Sat. Liquid h_f	Evap. h_{fg}	Sat. Vapor h_g	Sat. Liquid s_f	Sat. Vapor s_g	
−40	0.6417	0.6595	0.24191	−0.04	154.07	0.00	169.59	169.59	0.0000	0.7274	−40
−35	0.8071	0.6656	0.19540	4.37	156.13	4.42	167.48	171.90	0.0187	0.7219	−35
−30	1.0041	0.6720	0.15938	8.79	158.20	8.86	165.33	174.20	0.0371	0.7170	−30
−28	1.0927	0.6746	0.14728	10.58	159.02	10.65	164.46	175.11	0.0444	0.7153	−28
−26	1.1872	0.6773	0.13628	12.35	159.84	12.43	163.59	176.02	0.0517	0.7135	−26
−25	1.2368	0.6786	0.13117	13.25	160.26	13.33	163.15	176.48	0.0552	0.7126	−25
−24	1.2880	0.6800	0.12628	14.13	160.67	14.22	162.71	176.93	0.0589	0.7119	−24
−22	1.3953	0.6827	0.11717	15.92	161.48	16.02	161.82	177.83	0.0660	0.7103	−22
−20	1.5093	0.6855	0.10885	17.72	162.31	17.82	160.92	178.74	0.0731	0.7087	−20
−18	1.6304	0.6883	0.10124	19.51	163.12	19.62	160.01	179.63	0.0802	0.7073	−18
−15	1.8260	0.6926	0.09102	22.20	164.35	22.33	158.64	180.97	0.0906	0.7051	−15
−10	2.1912	0.7000	0.07665	26.72	166.39	26.87	156.31	183.19	0.1080	0.7019	−10
−5	2.6096	0.7078	0.06496	31.27	168.42	31.45	153.93	185.37	0.1251	0.6991	−5
0	3.0861	0.7159	0.05539	35.83	170.44	36.05	151.48	187.53	0.1420	0.6965	0
4	3.5124	0.7227	0.04895	39.51	172.04	39.76	149.47	189.23	0.1553	0.6946	4
8	3.9815	0.7297	0.04340	43.21	173.63	43.50	147.41	190.91	0.1686	0.6929	8
12	4.4962	0.7370	0.03860	46.93	175.20	47.26	145.30	192.56	0.1817	0.6913	12
16	5.0591	0.7446	0.03442	50.67	176.78	51.05	143.14	194.19	0.1948	0.6898	16
20	5.6729	0.7525	0.03078	54.44	178.32	54.87	140.91	195.78	0.2078	0.6884	20
24	6.3405	0.7607	0.02759	58.25	179.85	58.73	138.61	197.34	0.2207	0.6871	24
26	6.6954	0.7650	0.02614	60.17	180.61	60.68	137.44	198.11	0.2271	0.6865	26
28	7.0648	0.7694	0.02478	62.09	181.36	62.63	136.24	198.87	0.2335	0.6859	28
30	7.4490	0.7739	0.02351	64.01	182.11	64.59	135.03	199.62	0.2400	0.6853	30
32	7.8485	0.7785	0.02231	65.96	182.85	66.57	133.79	200.36	0.2463	0.6847	32
34	8.2636	0.7832	0.02118	67.90	183.59	68.55	132.53	201.09	0.2527	0.6842	34
36	8.6948	0.7880	0.02012	69.86	184.31	70.55	131.25	201.80	0.2591	0.6836	36
38	9.1423	0.7929	0.01912	71.84	185.03	72.56	129.94	202.51	0.2655	0.6831	38
40	9.6065	0.7980	0.01817	73.82	185.74	74.59	128.61	203.20	0.2718	0.6825	40
42	10.088	0.8033	0.01728	75.82	186.45	76.63	127.25	203.88	0.2782	0.6820	42
44	10.587	0.8086	0.01644	77.82	187.13	78.68	125.87	204.54	0.2845	0.6814	44
48	11.639	0.8199	0.01488	81.88	188.51	82.83	123.00	205.83	0.2973	0.6802	48
52	12.766	0.8318	0.01349	86.00	189.83	87.06	119.99	207.05	0.3101	0.6791	52
56	13.972	0.8445	0.01224	90.18	191.10	91.36	116.84	208.20	0.3229	0.6779	56
60	15.259	0.8581	0.01111	94.43	192.31	95.74	113.52	209.26	0.3358	0.6765	60
112	41.155	1.792	0.00179	175.98	175.98	183.35	0	183.35	0.5687	0.5687	112

Source: Tables A-7 through A-9 are adapted from K. Wark, *Thermodynamics, 4th ed.,* McGraw-Hill, New York, 1983, as based on data from Freon Products Division, E. I. du Pont de Nemours & Company, 1969.

TABLE A-8 Properties of Saturated Refrigerant 12 (Liquid–Vapor): Pressure Table

Press. bars	Temp. °C	Specific Volume m³/kg		Internal Energy kJ/kg		Enthalpy kJ/kg			Entropy kJ/kg · K		Press. bars
		Sat. Liquid $v_f \times 10^3$	Sat. Vapor v_g	Sat. Liquid u_f	Sat. Vapor u_g	Sat. Liquid h_f	Evap. h_{fg}	Sat. Vapor h_g	Sat. Liquid s_f	Sat. Vapor s_g	
0.6	−41.42	0.6578	0.2575	−1.29	153.49	−1.25	170.19	168.94	−0.0054	0.7290	0.6
1.0	−30.10	0.6719	0.1600	8.71	158.15	8.78	165.37	174.15	0.0368	0.7171	1.0
1.2	−25.74	0.6776	0.1349	12.58	159.95	12.66	163.48	176.14	0.0526	0.7133	1.2
1.4	−21.91	0.6828	0.1168	15.99	161.52	16.09	161.78	177.87	0.0663	0.7102	1.4
1.6	−18.49	0.6876	0.1031	19.07	162.91	19.18	160.23	179.41	0.0784	0.7076	1.6
1.8	−15.38	0.6921	0.09225	21.86	164.19	21.98	158.82	180.80	0.0893	0.7054	1.8
2.0	−12.53	0.6962	0.08354	24.43	165.36	24.57	157.50	182.07	0.0992	0.7035	2.0
2.4	−7.42	0.7040	0.07033	29.06	167.44	29.23	155.09	184.32	0.1168	0.7004	2.4
2.8	−2.93	0.7111	0.06076	33.15	169.26	33.35	152.92	186.27	0.1321	0.6980	2.8
3.2	1.11	0.7177	0.05351	36.85	170.88	37.08	150.92	188.00	0.1457	0.6960	3.2
4.0	8.15	0.7299	0.04321	43.35	173.69	43.64	147.33	190.97	0.1691	0.6928	4.0
5.0	15.60	0.7438	0.03482	50.30	176.61	50.67	143.35	194.02	0.1935	0.6899	5.0
6.0	22.00	0.7566	0.02913	56.35	179.09	56.80	139.77	196.57	0.2142	0.6878	6.0
7.0	27.65	0.7686	0.02501	61.75	181.23	62.29	136.45	198.74	0.2324	0.6860	7.0
8.0	32.74	0.7802	0.02188	66.68	183.13	67.30	133.33	200.63	0.2487	0.6845	8.0
9.0	37.37	0.7914	0.01942	71.22	184.81	71.93	130.36	202.29	0.2634	0.6832	9.0
10.0	41.64	0.8023	0.01744	75.46	186.32	76.26	127.50	203.76	0.2770	0.6820	10.0
12.0	49.31	0.8237	0.01441	83.22	188.95	84.21	122.03	206.24	0.3015	0.6799	12.0
14.0	56.09	0.8448	0.01222	90.28	191.11	91.46	116.76	208.22	0.3232	0.6778	14.0
16.0	62.19	0.8660	0.01054	96.80	192.95	98.19	111.62	209.81	0.3329	0.6758	16.0

TABLE A-9 Properties of Superheated Refrigerant 12

T °C	v m³/kg	u kJ/kg	h kJ/kg	s kJ/kg · K	v m³/kg	u kJ/kg	h kJ/kg	s kJ/kg · K
	$p = 0.6$ bar $= 0.06$ MPa ($T_{sat} = -41.42°C$)				$p = 1.0$ bar $= 0.10$ MPa ($T_{sat} = -30.10°C$)			
Sat.	0.2575	153.49	168.94	0.7290	0.1600	158.15	174.15	0.7171
−40	0.2593	154.16	169.72	0.7324				
−20	0.2838	163.91	180.94	0.7785	0.1677	163.22	179.99	0.7406
0	0.3079	174.05	192.52	0.8225	0.1827	173.50	191.77	0.7854
10	0.3198	179.26	198.45	0.8439	0.1900	178.77	197.77	0.8070
20	0.3317	184.57	204.47	0.8647	0.1973	184.12	203.85	0.8281
30	0.3435	189.96	210.57	0.8852	0.2045	189.57	210.02	0.8488
40	0.3552	195.46	216.77	0.9053	0.2117	195.09	216.26	0.8691
50	0.3670	201.02	223.04	0.9251	0.2188	200.70	222.58	0.8889
60	0.3787	206.69	229.41	0.9444	0.2260	206.38	228.98	0.9084
80	0.4020	218.25	242.37	0.9822	0.2401	218.00	242.01	0.9464
	$p = 1.4$ bars $= 0.14$ MPa ($T_{sat} = -21.91°C$)				$p = 1.8$ bars $= 0.18$ MPa ($T_{sat} = -15.38°C$)			
Sat.	0.1168	161.52	177.87	0.7102	0.0922	164.20	180.80	0.7054
−20	0.1179	162.50	179.01	0.7147				
−10	0.1235	167.69	184.97	0.7378	0.0925	164.39	181.03	0.7181
0	0.1289	172.94	190.99	0.7602	0.0991	172.37	190.21	0.7408
10	0.1343	178.28	197.08	0.7821	0.1034	177.77	196.38	0.7630
20	0.3197	183.67	203.23	0.8035	0.1076	183.23	202.60	0.7846
30	0.1449	189.17	209.46	0.8243	0.1118	188.77	208.89	0.8057
40	0.1502	194.72	215.75	0.8447	0.1160	194.35	215.23	0.8263
50	0.1553	200.38	222.12	0.8648	0.1201	200.02	221.64	0.8464
60	0.1605	206.08	228.55	0.8844	0.1241	205.78	228.12	0.8662
80	0.1707	217.74	241.64	0.9225	0.1322	217.47	241.27	0.9045
100	0.1809	229.67	255.00	0.9593	0.1402	229.45	254.69	0.9414
	$p = 2.0$ bars $= 0.20$ MPa ($T_{sat} = -12.53°C$)				$p = 2.4$ bars $= 0.24$ MPa ($T_{sat} = -7.42°C$)			
Sat.	0.0835	165.37	182.07	0.7035	0.0703	167.45	184.32	0.7004
0	0.0886	172.08	189.08	0.7325	0.0729	171.49	188.99	0.7177
10	0.0926	177.50	196.02	0.7548	0.0763	176.98	195.29	0.7404
20	0.0964	183.00	202.28	0.7766	0.0796	182.53	201.63	0.7624
30	0.1002	188.56	208.60	0.7978	0.0828	188.14	208.01	0.7838
40	0.1040	194.17	214.97	0.8184	0.0860	193.80	214.44	0.8047
50	0.1077	199.86	221.40	0.8387	0.0892	199.51	220.92	0.8251
60	0.1114	205.62	227.90	0.8585	0.0923	205.31	227.46	0.8450
80	0.1187	217.35	241.09	0.8969	0.0985	217.07	240.71	0.8836
100	0.1259	229.35	254.53	0.9339	0.1045	229.12	254.20	0.9208
120	0.1331	241.59	268.21	0.9696	0.1105	241.41	267.93	0.9566

TABLE A-9 (*Continued*)

T °C	v m³/kg	u kJ/kg	h kJ/kg	s kJ/kg · K	v m³/kg	u kJ/kg	h kJ/kg	s kJ/kg · K
	p = 2.8 bars = 0.28 MPa (T_{sat} = −2.93°C)				*p* = 3.2 bars = 0.32 MPa (T_{sat} = 1.11°C)			
Sat.	0.06076	169.26	186.27	0.6980	0.05351	170.88	188.00	0.6960
0	0.06166	170.89	188.15	0.7049				
10	0.06464	176.45	194.55	0.7279	0.05590	175.90	193.79	0.7167
20	0.06755	182.06	200.97	0.7502	0.05852	181.57	200.30	0.7393
30	0.07040	187.71	207.42	0.7718	0.06106	187.28	206.82	0.7612
40	0.07319	193.42	213.91	0.7928	0.06355	193.02	213.36	0.7824
50	0.07594	199.18	220.44	0.8134	0.06600	198.82	219.94	0.8031
60	0.07865	205.00	227.02	0.8334	0.06841	204.68	226.57	0.8233
80	0.08399	216.82	240.34	0.8722	0.07314	216.55	239.96	0.8623
100	0.08924	228.29	253.88	0.9095	0.07778	228.66	253.55	0.8997
120	0.09443	241.21	267.65	0.9455	0.08236	241.00	267.36	0.9358
	p = 4.0 bars = 0.40 MPa (T_{sat} = 8.15°C)				*p* = 5.0 bars = 0.50 MPa (T_{sat} = 15.60°C)			
Sat.	0.04321	173.69	190.97	0.6928	0.03482	176.61	194.02	0.6899
10	0.04363	174.76	192.21	0.6972				
20	0.04584	180.57	198.91	0.7204	0.03565	179.26	197.08	0.7004
30	0.04797	186.39	205.58	0.7428	0.03746	185.23	203.96	0.7235
40	0.05005	192.23	212.25	0.7645	0.03922	191.20	210.81	0.7457
50	0.05207	198.11	218.94	0.7855	0.04091	197.19	217.64	0.7672
60	0.05406	204.03	225.65	0.8060	0.04257	203.20	224.48	0.7881
80	0.05791	216.03	239.19	0.8454	0.04578	215.32	238.21	0.8281
100	0.06173	228.20	252.89	0.8831	0.04889	227.61	252.05	0.8662
120	0.06546	240.61	266.79	0.9194	0.05193	240.10	266.06	0.9028
140	0.06913	253.23	280.88	0.9544	0.05492	252.77	280.23	0.9379
	p = 6.0 bars = 0.60 MPa (T_{sat} = 22.00°C)				*p* = 7.0 bars = 0.70 MPa (T_{sat} = 27.65°C)			
Sat.	0.02913	179.09	196.57	0.6878	0.02501	181.23	198.74	0.6860
30	0.03042	184.01	202.26	0.7068	0.02535	182.72	200.46	0.6917
40	0.03197	190.13	209.31	0.7297	0.02676	189.00	207.73	0.7153
50	0.03345	196.23	216.30	0.7516	0.02810	195.23	214.90	0.7378
60	0.03489	202.34	223.27	0.7729	0.02939	201.45	222.02	0.7595
80	0.03765	214.61	237.20	0.8135	0.03184	213.88	236.17	0.8008
100	0.04032	227.01	251.20	0.8520	0.03419	226.40	250.33	0.8398
120	0.04291	239.57	265.32	0.8889	0.03646	239.05	264.57	0.8769
140	0.04545	252.31	279.58	0.9243	0.03867	251.85	278.92	0.9125
160	0.04794	265.25	294.01	0.9584	0.04085	264.83	293.42	0.9468

TABLE A-9 (*Continued*)

T °C	v m³/kg	u kJ/kg	h kJ/kg	s kJ/kg · K	v m³/kg	u kJ/kg	h kJ/kg	s kJ/kg · K
	p = 8.0 bars = 0.80 MPa (*T*ₛₐₜ = 32.74°C)				*p* = 9.0 bars = 0.90 MPa (*T*ₛₐₜ = 37.37°C)			
Sat.	0.02188	183.13	200.63	0.6845	0.01942	184.81	202.29	0.6832
40	0.02283	187.81	206.07	0.7021	0.01974	186.55	204.32	0.6897
50	0.02407	194.19	213.45	0.7253	0.02091	193.10	211.92	0.7136
60	0.02525	200.52	220.72	0.7474	0.02201	199.56	219.37	0.7363
80	0.02748	213.13	235.11	0.7894	0.02407	212.37	234.03	0.7790
100	0.02959	225.77	249.44	0.8289	0.02601	225.13	248.54	0.8190
120	0.03162	238.51	263.81	0.8664	0.02785	237.97	263.03	0.8569
140	0.03359	251.39	278.26	0.9022	0.02964	250.90	277.58	0.8930
160	0.03552	264.41	292.83	0.9367	0.03138	263.99	292.23	0.9276
180	0.03742	277.60	307.54	0.9699	0.03309	277.23	307.01	0.9609
	p = 10.0 bars = 1.00 MPa (*T*ₛₐₜ = 41.64°C)				*p* = 12.0 bars = 1.20 MPa (*T*ₛₐₜ = 49.31°C)			
Sat.	0.01744	186.32	203.76	0.6820	0.01441	188.95	206.24	0.6799
50	0.01837	191.95	210.32	0.7026	0.01448	189.43	206.81	0.6816
60	0.01941	198.56	217.97	0.7259	0.01546	196.41	214.96	0.7065
80	0.02134	211.57	232.91	0.7695	0.01722	209.91	230.57	0.7520
100	0.02313	224.48	247.61	0.8100	0.01881	223.13	245.70	0.7937
120	0.02484	237.41	262.25	0.8482	0.02030	236.27	260.63	0.8326
140	0.02647	250.43	276.90	0.8845	0.02172	249.45	275.51	0.8696
160	0.02807	263.56	291.63	0.9193	0.02309	263.70	290.41	0.9048
180	0.02963	276.84	306.47	0.9528	0.02443	276.05	305.37	0.9385
200	0.03116	290.26	321.42	0.9851	0.02574	289.55	320.44	0.9711
	p = 14.0 bars = 1.40 MPa (*T*ₛₐₜ = 56.09°C)				*p* = 16.0 bars = 1.60 MPa (*T*ₛₐₜ = 62.19°C)			
Sat.	0.01222	191.11	208.22	0.6778	0.01054	192.95	209.81	0.6758
60	0.01258	194.00	211.61	0.6881				
80	0.01425	208.11	228.06	0.7360	0.01198	206.17	225.34	0.7209
100	0.01571	221.70	243.69	0.7791	0.01337	220.19	241.58	0.7656
120	0.01705	235.09	258.96	0.8189	0.01461	233.84	257.22	0.8065
140	0.01832	248.43	274.08	0.8564	0.01577	247.38	272.61	0.8447
160	0.01954	261.80	289.16	0.8921	0.01686	260.90	287.88	0.8808
180	0.02071	275.27	304.26	0.9262	0.01792	274.47	303.14	0.9152
200	0.02186	288.84	319.44	0.9589	0.01895	288.11	318.43	0.9482
220	0.02299	302.51	334.70	0.9905	0.01996	301.84	333.78	0.9800

TABLE A-10 Properties of Saturated Refrigerant 134a (Liquid–Vapor): Temperature Table

Temp. °C	Press. bars	Specific Volume m³/kg		Internal Energy kJ/kg		Enthalpy kJ/kg			Entropy kJ/kg · K		Temp. °C
		Sat. Liquid $v_f \times 10^3$	Sat. Vapor v_g	Sat. Liquid u_f	Sat. Vapor u_g	Sat. Liquid h_f	Evap. h_{fg}	Sat. Vapor h_g	Sat. Liquid s_f	Sat. Vapor s_g	
−40	0.5164	0.7055	0.3569	−0.04	204.45	0.00	222.88	222.88	0.0000	0.9560	−40
−36	0.6332	0.7113	0.2947	4.68	206.73	4.73	220.67	225.40	0.0201	0.9506	−36
−32	0.7704	0.7172	0.2451	9.47	209.01	9.52	218.37	227.90	0.0401	0.9456	−32
−28	0.9305	0.7233	0.2052	14.31	211.29	14.37	216.01	230.38	0.0600	0.9411	−28
−26	1.0199	0.7265	0.1882	16.75	212.43	16.82	214.80	231.62	0.0699	0.9390	−26
−24	1.1160	0.7296	0.1728	19.21	213.57	19.29	213.57	232.85	0.0798	0.9370	−24
−22	1.2192	0.7328	0.1590	21.68	214.70	21.77	212.32	234.08	0.0897	0.9351	−22
−20	1.3299	0.7361	0.1464	24.17	215.84	24.26	211.05	235.31	0.0996	0.9332	−20
−18	1.4483	0.7395	0.1350	26.67	216.97	26.77	209.76	236.53	0.1094	0.9315	−18
−16	1.5748	0.7428	0.1247	29.18	218.10	29.30	208.45	237.74	0.1192	0.9298	−16
−12	1.8540	0.7498	0.1068	34.25	220.36	34.39	205.77	240.15	0.1388	0.9267	−12
−8	2.1704	0.7569	0.0919	39.38	222.60	39.54	203.00	242.54	0.1583	0.9239	−8
−4	2.5274	0.7644	0.0794	44.56	224.84	44.75	200.15	244.90	0.1777	0.9213	−4
0	2.9282	0.7721	0.0689	49.79	227.06	50.02	197.21	247.23	0.1970	0.9190	0
4	3.3765	0.7801	0.0600	55.08	229.27	55.35	194.19	249.53	0.2162	0.9169	4
8	3.8756	0.7884	0.0525	60.43	231.46	60.73	191.07	251.80	0.2354	0.9150	8
12	4.4294	0.7971	0.0460	65.83	233.63	66.18	187.85	254.03	0.2545	0.9132	12
16	5.0416	0.8062	0.0405	71.29	235.78	71.69	184.52	256.22	0.2735	0.9116	16
20	5.7160	0.8157	0.0358	76.80	237.91	77.26	181.09	258.36	0.2924	0.9102	20
24	6.4566	0.8257	0.0317	82.37	240.01	82.90	177.55	260.45	0.3113	0.9089	24
26	6.8530	0.8309	0.0298	85.18	241.05	85.75	175.73	261.48	0.3208	0.9082	26
28	7.2675	0.8362	0.0281	88.00	242.08	88.61	173.89	262.50	0.3302	0.9076	28
30	7.7006	0.8417	0.0265	90.84	243.10	91.49	172.00	263.50	0.3396	0.9070	30
32	8.1528	0.8473	0.0250	93.70	244.12	94.39	170.09	264.48	0.3490	0.9064	32
34	8.6247	0.8530	0.0236	96.58	245.12	97.31	168.14	265.45	0.3584	0.9058	34
36	9.1168	0.8590	0.0223	99.47	246.11	100.25	166.15	266.40	0.3678	0.9053	36
38	9.6298	0.8651	0.0210	102.38	247.09	103.21	164.12	267.33	0.3772	0.9047	38
40	10.164	0.8714	0.0199	105.30	248.06	106.19	162.05	268.24	0.3866	0.9041	40
42	10.720	0.8780	0.0188	108.25	249.02	109.19	159.94	269.14	0.3960	0.9035	42
44	11.299	0.8847	0.0177	111.22	249.96	112.22	157.79	270.01	0.4054	0.9030	44
48	12.526	0.8989	0.0159	117.22	251.79	118.35	153.33	271.68	0.4243	0.9017	48
52	13.851	0.9142	0.0142	123.31	253.55	124.58	148.66	273.24	0.4432	0.9004	52
56	15.278	0.9308	0.0127	129.51	255.23	130.93	143.75	274.68	0.4622	0.8990	56
60	16.813	0.9488	0.0114	135.82	256.81	137.42	138.57	275.99	0.4814	0.8973	60
70	21.162	1.0027	0.0086	152.22	260.15	154.34	124.08	278.43	0.5302	0.8918	70
80	26.324	1.0766	0.0064	169.88	262.14	172.71	106.41	279.12	0.5814	0.8827	80
90	32.435	1.1949	0.0046	189.82	261.34	193.69	82.63	276.32	0.6380	0.8655	90
100	39.742	1.5443	0.0027	218.60	248.49	224.74	34.40	259.13	0.7196	0.8117	100

Source: Tables A-10 through A-12 are calculated based on equations from D. P. Wilson and R. S. Basu, "Thermodynamic Properties of a New Stratospherically Safe Working Fluid—Refrigerant 134a," *ASHRAE Trans.*, Vol. 94, Pt. 2, 1988, pp. 2095–2118.

TABLE A-11 Properties of Saturated Refrigerant 134a (Liquid–Vapor): Pressure Table

Press. bars	Temp. °C	Specific Volume m³/kg		Internal Energy kJ/kg		Enthalpy kJ/kg			Entropy kJ/kg · K		Press. bars
		Sat. Liquid $v_f \times 10^3$	Sat. Vapor v_g	Sat. Liquid u_f	Sat. Vapor u_g	Sat. Liquid h_f	Evap. h_{fg}	Sat. Vapor h_g	Sat. Liquid s_f	Sat. Vapor s_g	
0.6	−37.07	0.7097	0.3100	3.41	206.12	3.46	221.27	224.72	0.0147	0.9520	0.6
0.8	−31.21	0.7184	0.2366	10.41	209.46	10.47	217.92	228.39	0.0440	0.9447	0.8
1.0	−26.43	0.7258	0.1917	16.22	212.18	16.29	215.06	231.35	0.0678	0.9395	1.0
1.2	−22.36	0.7323	0.1614	21.23	214.50	21.32	212.54	233.86	0.0879	0.9354	1.2
1.4	−18.80	0.7381	0.1395	25.66	216.52	25.77	210.27	236.04	0.1055	0.9322	1.4
1.6	−15.62	0.7435	0.1229	29.66	218.32	29.78	208.19	237.97	0.1211	0.9295	1.6
1.8	−12.73	0.7485	0.1098	33.31	219.94	33.45	206.26	239.71	0.1352	0.9273	1.8
2.0	−10.09	0.7532	0.0993	36.69	221.43	36.84	204.46	241.30	0.1481	0.9253	2.0
2.4	−5.37	0.7618	0.0834	42.77	224.07	42.95	201.14	244.09	0.1710	0.9222	2.4
2.8	−1.23	0.7697	0.0719	48.18	226.38	48.39	198.13	246.52	0.1911	0.9197	2.8
3.2	2.48	0.7770	0.0632	53.06	228.43	53.31	195.35	248.66	0.2089	0.9177	3.2
3.6	5.84	0.7839	0.0564	57.54	230.28	57.82	192.76	250.58	0.2251	0.9160	3.6
4.0	8.93	0.7904	0.0509	61.69	231.97	62.00	190.32	252.32	0.2399	0.9145	4.0
5.0	15.74	0.8056	0.0409	70.93	235.64	71.33	184.74	256.07	0.2723	0.9117	5.0
6.0	21.58	0.8196	0.0341	78.99	238.74	79.48	179.71	259.19	0.2999	0.9097	6.0
7.0	26.72	0.8328	0.0292	86.19	241.42	86.78	175.07	261.85	0.3242	0.9080	7.0
8.0	31.33	0.8454	0.0255	92.75	243.78	93.42	170.73	264.15	0.3459	0.9066	8.0
9.0	35.53	0.8576	0.0226	98.79	245.88	99.56	166.62	266.18	0.3656	0.9054	9.0
10.0	39.39	0.8695	0.0202	104.42	247.77	105.29	162.68	267.97	0.3838	0.9043	10.0
12.0	46.32	0.8928	0.0166	114.69	251.03	115.76	155.23	270.99	0.4164	0.9023	12.0
14.0	52.43	0.9159	0.0140	123.98	253.74	125.26	148.14	273.40	0.4453	0.9003	14.0
16.0	57.92	0.9392	0.0121	132.52	256.00	134.02	141.31	275.33	0.4714	0.8982	16.0
18.0	62.91	0.9631	0.0105	140.49	257.88	142.22	134.60	276.83	0.4954	0.8959	18.0
20.0	67.49	0.9878	0.0093	148.02	259.41	149.99	127.95	277.94	0.5178	0.8934	20.0
25.0	77.59	1.0562	0.0069	165.48	261.84	168.12	111.06	279.17	0.5687	0.8854	25.0
30.0	86.22	1.1416	0.0053	181.88	262.16	185.30	92.71	278.01	0.6156	0.8735	30.0

TABLE A-12 Properties of Superheated Refrigerant 134a

T °C	v m³/kg	u kJ/kg	h kJ/kg	s kJ/kg · K	v m³/kg	u kJ/kg	h kJ/kg	s kJ/kg · K
	p = 0.6 bars = 0.06 MPa (T_{sat} = −37.07°C)				**p = 1.0 bars = 0.10 MPa** (T_{sat} = −26.43°C)			
Sat.	0.31003	206.12	224.72	0.9520	0.19170	212.18	231.35	0.9395
−20	0.33536	217.86	237.98	1.0062	0.19770	216.77	236.54	0.9602
−10	0.34992	224.97	245.96	1.0371	0.20686	224.01	244.70	0.9918
0	0.36433	232.24	254.10	1.0675	0.21587	231.41	252.99	1.0227
10	0.37861	239.69	262.41	1.0973	0.22473	238.96	261.43	1.0531
20	0.39279	247.32	270.89	1.1267	0.23349	246.67	270.02	1.0829
30	0.40688	255.12	279.53	1.1557	0.24216	254.54	278.76	1.1122
40	0.42091	263.10	288.35	1.1844	0.25076	262.58	287.66	1.1411
50	0.43487	271.25	297.34	1.2126	0.25930	270.79	296.72	1.1696
60	0.44879	279.58	306.51	1.2405	0.26779	279.16	305.94	1.1977
70	0.46266	288.08	315.84	1.2681	0.27623	287.70	315.32	1.2254
80	0.47650	296.75	325.34	1.2954	0.28464	296.40	324.87	1.2528
90	0.49031	305.58	335.00	1.3224	0.29302	305.27	334.57	1.2799
	p = 1.4 bars = 0.14 MPa (T_{sat} = −18.80°C)				**p = 1.8 bars = 0.18 MPa** (T_{sat} = −12.73°C)			
Sat.	0.13945	216.52	236.04	0.9322	0.10983	219.94	239.71	0.9273
−10	0.14549	223.03	243.40	0.9606	0.11135	222.02	242.06	0.9362
0	0.15219	230.55	251.86	0.9922	0.11678	229.67	250.69	0.9684
10	0.15875	238.21	260.43	1.0230	0.12207	237.44	259.41	0.9998
20	0.16520	246.01	269.13	1.0532	0.12723	245.33	268.23	1.0304
30	0.17155	253.96	277.97	1.0828	0.13230	253.36	277.17	1.0604
40	0.17783	262.06	286.96	1.1120	0.13730	261.53	286.24	1.0898
50	0.18404	270.32	296.09	1.1407	0.14222	269.85	295.45	1.1187
60	0.19020	278.74	305.37	1.1690	0.14710	278.31	304.79	1.1472
70	0.19633	287.32	314.80	1.1969	0.15193	286.93	314.28	1.1753
80	0.20241	296.06	324.39	1.2244	0.15672	295.71	323.92	1.2030
90	0.20846	304.95	334.14	1.2516	0.16148	304.63	333.70	1.2303
100	0.21449	314.01	344.04	1.2785	0.16622	313.72	343.63	1.2573
	p = 2.0 bars = 0.20 MPa (T_{sat} = −10.09°C)				**p = 2.4 bars = 0.24 MPa** (T_{sat} = −5.37°C)			
Sat.	0.09933	221.43	241.30	0.9253	0.08343	224.07	244.09	0.9222
−10	0.09938	221.50	241.38	0.9256				
0	0.10438	229.23	250.10	0.9582	0.08574	228.31	248.89	0.9399
10	0.10922	237.05	258.89	0.9898	0.08993	236.26	257.84	0.9721
20	0.11394	244.99	267.78	1.0206	0.09399	244.30	266.85	1.0034
30	0.11856	253.06	276.77	1.0508	0.09794	252.45	275.95	1.0339
40	0.12311	261.26	285.88	1.0804	0.10181	260.72	285.16	1.0637
50	0.12758	269.61	295.12	1.1094	0.10562	269.12	294.47	1.0930
60	0.13201	278.10	304.50	1.1380	0.10937	277.67	303.91	1.1218
70	0.13639	286.74	314.02	1.1661	0.11307	286.35	313.49	1.1501
80	0.14073	295.53	323.68	1.1939	0.11674	295.18	323.19	1.1780
90	0.14504	304.47	333.48	1.2212	0.12037	304.15	333.04	1.2055
100	0.14932	313.57	343.43	1.2483	0.12398	313.27	343.03	1.2326

TABLE A-12 (*Continued*)

T °C	v m³/kg	u kJ/kg	h kJ/kg	s kJ/kg · K	v m³/kg	u kJ/kg	h kJ/kg	s kJ/kg · K
	p = 2.8 bars = 0.28 MPa (*T*ₛₐₜ = −1.23°C)				*p* = 3.2 bars = 0.32 MPa (*T*ₛₐₜ = 2.48°C)			
Sat.	0.07193	226.38	246.52	0.9197	0.06322	228.43	248.66	0.9177
0	0.07240	227.37	247.64	0.9238				
10	0.07613	235.44	256.76	0.9566	0.06576	234.61	255.65	0.9427
20	0.07972	243.59	265.91	0.9883	0.06901	242.87	264.95	0.9749
30	0.08320	251.83	275.12	1.0192	0.07214	251.19	274.28	1.0062
40	0.08660	260.17	284.42	1.0494	0.07518	259.61	283.67	1.0367
50	0.08992	268.64	293.81	1.0789	0.07815	268.14	293.15	1.0665
60	0.09319	277.23	303.32	1.1079	0.08106	276.79	302.72	1.0957
70	0.09641	285.96	312.95	1.1364	0.08392	285.56	312.41	1.1243
80	0.09960	294.82	322.71	1.1644	0.08674	294.46	322.22	1.1525
90	0.10275	303.83	332.60	1.1920	0.08953	303.50	332.15	1.1802
100	0.10587	312.98	342.62	1.2193	0.09229	312.68	342.21	1.2076
110	0.10897	322.27	352.78	1.2461	0.09503	322.00	352.40	1.2345
120	0.11205	331.71	363.08	1.2727	0.09774	331.45	362.73	1.2611

T °C	v m³/kg	u kJ/kg	h kJ/kg	s kJ/kg · K	v m³/kg	u kJ/kg	h kJ/kg	s kJ/kg · K
	p = 4.0 bars = 0.40 MPa (*T*ₛₐₜ = 8.93°C)				*p* = 5.0 bars = 0.50 MPa (*T*ₛₐₜ = 15.74°C)			
Sat.	0.05089	231.97	252.32	0.9145	0.04086	235.64	256.07	0.9117
10	0.05119	232.87	253.35	0.9182				
20	0.05397	241.37	262.96	0.9515	0.04188	239.40	260.34	0.9264
30	0.05662	249.89	272.54	0.9837	0.04416	248.20	270.28	0.9597
40	0.05917	258.47	282.14	1.0148	0.04633	256.99	280.16	0.9918
50	0.06164	267.13	291.79	1.0452	0.04842	265.83	290.04	1.0229
60	0.06405	275.89	301.51	1.0748	0.05043	274.73	299.95	1.0531
70	0.06641	284.75	311.32	1.1038	0.05240	283.72	309.92	1.0825
80	0.06873	293.73	321.23	1.1322	0.05432	292.80	319.96	1.1114
90	0.07102	302.84	331.25	1.1602	0.05620	302.00	330.10	1.1397
100	0.07327	312.07	341.38	1.1878	0.05805	311.31	340.33	1.1675
110	0.07550	321.44	351.64	1.2149	0.05988	320.74	350.68	1.1949
120	0.07771	330.94	362.03	1.2417	0.06168	330.30	361.14	1.2218
130	0.07991	340.58	372.54	1.2681	0.06347	339.98	371.72	1.2484
140	0.08208	350.35	383.18	1.2941	0.06524	349.79	382.42	1.2746

TABLE A-12 (*Continued*)

T °C	v m³/kg	u kJ/kg	h kJ/kg	s kJ/kg · K	v m³/kg	u kJ/kg	h kJ/kg	s kJ/kg · K
	p = 6.0 bars = 0.60 MPa (T_{sat} = 21.58°C)				p = 7.0 bars = 0.70 MPa (T_{sat} = 26.72°C)			
Sat.	0.03408	238.74	259.19	0.9097	0.02918	241.42	261.85	0.9080
30	0.03581	246.41	267.89	0.9388	0.02979	244.51	265.37	0.9197
40	0.03774	255.45	278.09	0.9719	0.03157	253.83	275.93	0.9539
50	0.03958	264.48	288.23	1.0037	0.03324	263.08	286.35	0.9867
60	0.04134	273.54	298.35	1.0346	0.03482	272.31	296.69	1.0182
70	0.04304	282.66	308.48	1.0645	0.03634	281.57	307.01	1.0487
80	0.04469	291.86	318.67	1.0938	0.03781	290.88	317.35	1.0784
90	0.04631	301.14	328.93	1.1225	0.03924	300.27	327.74	1.1074
100	0.04790	310.53	339.27	1.1505	0.04064	309.74	338.19	1.1358
110	0.04946	320.03	349.70	1.1781	0.04201	319.31	348.71	1.1637
120	0.05099	329.64	360.24	1.2053	0.04335	328.98	359.33	1.1910
130	0.05251	339.38	370.88	1.2320	0.04468	338.76	370.04	1.2179
140	0.05402	349.23	381.64	1.2584	0.04599	348.66	380.86	1.2444
150	0.05550	359.21	392.52	1.2844	0.04729	358.68	391.79	1.2706
160	0.05698	369.32	403.51	1.3100	0.04857	368.82	402.82	1.2963
	p = 8.0 bars = 0.80 MPa (T_{sat} = 31.33°C)				p = 9.0 bars = 0.90 MPa (T_{sat} = 35.53°C)			
Sat.	0.02547	243.78	264.15	0.9066	0.02255	245.88	266.18	0.9054
40	0.02691	252.13	273.66	0.9374	0.02325	250.32	271.25	0.9217
50	0.02846	261.62	284.39	0.9711	0.02472	260.09	282.34	0.9566
60	0.02992	271.04	294.98	1.0034	0.02609	269.72	293.21	0.9897
70	0.03131	280.45	305.50	1.0345	0.02738	279.30	303.94	1.0214
80	0.03264	289.89	316.00	1.0647	0.02861	288.87	314.62	1.0521
90	0.03393	299.37	326.52	1.0940	0.02980	298.46	325.28	1.0819
100	0.03519	308.93	337.08	1.1227	0.03095	308.11	335.96	1.1109
110	0.03642	318.57	347.71	1.1508	0.03207	317.82	346.68	1.1392
120	0.03762	328.31	358.40	1.1784	0.03316	327.62	357.47	1.1670
130	0.03881	338.14	369.19	1.2055	0.03423	337.52	368.33	1.1943
140	0.03997	348.09	380.07	1.2321	0.03529	347.51	379.27	1.2211
150	0.04113	358.15	391.05	1.2584	0.03633	357.61	390.31	1.2475
160	0.04227	368.32	402.14	1.2843	0.03736	367.82	401.44	1.2735
170	0.04340	378.61	413.33	1.3098	0.03838	378.14	412.68	1.2992
180	0.04452	389.02	424.63	1.3351	0.03939	388.57	424.02	1.3245

TABLE A-12 (*Continued*)

T °C	v m³/kg	u kJ/kg	h kJ/kg	s kJ/kg · K	v m³/kg	u kJ/kg	h kJ/kg	s kJ/kg · K
	p = 10.0 bars = 1.00 MPa (T_{sat} = 39.39°C)				*p* = 12.0 bars = 1.20 MPa (T_{sat} = 46.32°C)			
Sat.	0.02020	247.77	267.97	0.9043	0.01663	251.03	270.99	0.9023
40	0.02029	248.39	268.68	0.9066				
50	0.02171	258.48	280.19	0.9428	0.01712	254.98	275.52	0.9164
60	0.02301	268.35	291.36	0.9768	0.01835	265.42	287.44	0.9527
70	0.02423	278.11	302.34	1.0093	0.01947	275.59	298.96	0.9868
80	0.02538	287.82	313.20	1.0405	0.02051	285.62	310.24	1.0192
90	0.02649	297.53	324.01	1.0707	0.02150	295.59	321.39	1.0503
100	0.02755	307.27	334.82	1.1000	0.02244	305.54	332.47	1.0804
110	0.02858	317.06	345.65	1.1286	0.02335	315.50	343.52	1.1096
120	0.02959	326.93	356.52	1.1567	0.02423	325.51	354.58	1.1381
130	0.03058	336.88	367.46	1.1841	0.02508	335.58	365.68	1.1660
140	0.03154	346.92	378.46	1.2111	0.02592	345.73	376.83	1.1933
150	0.03250	357.06	389.56	1.2376	0.02674	355.95	388.04	1.2201
160	0.03344	367.31	400.74	1.2638	0.02754	366.27	399.33	1.2465
170	0.03436	377.66	412.02	1.2895	0.02834	376.69	410.70	1.2724
180	0.03528	388.12	423.40	1.3149	0.02912	387.21	422.16	1.2980

T °C	v m³/kg	u kJ/kg	h kJ/kg	s kJ/kg · K	v m³/kg	u kJ/kg	h kJ/kg	s kJ/kg · K
	p = 14.0 bars = 1.40 MPa (T_{sat} = 52.43°C)				*p* = 16.0 bars = 1.60 MPa (T_{sat} = 57.92°C)			
Sat.	0.01405	253.74	273.40	0.9003	0.01208	256.00	275.33	0.8982
60	0.01495	262.17	283.10	0.9297	0.01233	258.48	278.20	0.9069
70	0.01603	272.87	295.31	0.9658	0.01340	269.89	291.33	0.9457
80	0.01701	283.29	307.10	0.9997	0.01435	280.78	303.74	0.9813
90	0.01792	293.55	318.63	1.0319	0.01521	291.39	315.72	1.0148
100	0.01878	303.73	330.02	1.0628	0.01601	301.84	327.46	1.0467
110	0.01960	313.88	341.32	1.0927	0.01677	312.20	339.04	1.0773
120	0.02039	324.05	352.59	1.1218	0.01750	322.53	350.53	1.1069
130	0.02115	334.25	363.86	1.1501	0.01820	332.87	361.99	1.1357
140	0.02189	344.50	375.15	1.1777	0.01887	343.24	373.44	1.1638
150	0.02262	354.82	386.49	1.2048	0.01953	353.66	384.91	1.1912
160	0.02333	365.22	397.89	1.2315	0.02017	364.15	396.43	1.2181
170	0.02403	375.71	409.36	1.2576	0.02080	374.71	407.99	1.2445
180	0.02472	386.29	420.90	1.2834	0.02142	385.35	419.62	1.2704
190	0.02541	396.96	432.53	1.3088	0.02203	396.08	431.33	1.2960
200	0.02608	407.73	444.24	1.3338	0.02263	406.90	443.11	1.3212

TABLE A-13 Specific Heats of Selected Liquids and Solids

A. Liquid	State	c (kJ/kg $\cdot$ K)
Ammonia	sat. liq., $-20°C$	4.52
	sat. liq., $10°C$	4.67
	sat. liq., $50°C$	5.10
Ethyl alcohol	1 atm, $25°C$	2.43
Glycerin	1 atm, $10°C$	2.32
	1 atm, $50°C$	2.58
Mercury	1 atm, $10°C$	0.138
	1 atm, $315°C$	0.134
Refrigerant 12	sat. liq., $-20°C$	0.90
	sat. liq., $20°C$	0.96
Water	1 atm, $0°C$	4.217
	1 atm, $27°C$	4.179
	1 atm, $100°C$	4.218

B. Solid	Temperature (K)	c (kJ/kg $\cdot$ K)
Aluminum	300	0.903
Copper	300	0.385
	400	0.393
Ice	200	1.56
	240	1.86
	273	2.11
Iron	300	0.447
Lead	300	0.129
Silver	300	0.235

TABLE A-14 Ideal Gas Specific Heats of Some Common Gases (kJ/kg · K)

Temp. K	c_p	c_v	k	c_p	c_v	k	c_p	c_v	k	Temp. K
	Air			Nitrogen, N_2			Oxygen, O_2			
250	1.003	0.716	1.401	1.039	0.742	1.400	0.913	0.653	1.398	250
300	1.005	0.718	1.400	1.039	0.743	1.400	0.918	0.658	1.395	300
350	1.008	0.721	1.398	1.041	0.744	1.399	0.928	0.668	1.389	350
400	1.013	0.726	1.395	1.044	0.747	1.397	0.941	0.681	1.382	400
450	1.020	0.733	1.391	1.049	0.752	1.395	0.956	0.696	1.373	450
500	1.029	0.742	1.387	1.056	0.759	1.391	0.972	0.712	1.365	500
550	1.040	0.753	1.381	1.065	0.768	1.387	0.988	0.728	1.358	550
600	1.051	0.764	1.376	1.075	0.778	1.382	1.003	0.743	1.350	600
650	1.063	0.776	1.370	1.086	0.789	1.376	1.017	0.758	1.343	650
700	1.075	0.788	1.364	1.098	0.801	1.371	1.031	0.771	1.337	700
750	1.087	0.800	1.359	1.110	0.813	1.365	1.043	0.783	1.332	750
800	1.099	0.812	1.354	1.121	0.825	1.360	1.054	0.794	1.327	800
900	1.121	0.834	1.344	1.145	0.849	1.349	1.074	0.814	1.319	900
1000	1.142	0.855	1.336	1.167	0.870	1.341	1.090	0.830	1.313	1000

Temp. K	c_p	c_v	k	c_p	c_v	k	c_p	c_v	k	Temp. K
	Carbon dioxide, CO_2			Carbon monoxide, CO			Hydrogen, H_2			
250	0.791	0.602	1.314	1.039	0.743	1.400	14.051	9.927	1.416	250
300	0.846	0.657	1.288	1.040	0.744	1.399	14.307	10.183	1.405	300
350	0.895	0.706	1.268	1.043	0.746	1.398	14.427	10.302	1.400	350
400	0.939	0.750	1.252	1.047	0.751	1.395	14.476	10.352	1.398	400
450	0.978	0.790	1.239	1.054	0.757	1.392	14.501	10.377	1.398	450
500	1.014	0.825	1.229	1.063	0.767	1.387	14.513	10.389	1.397	500
550	1.046	0.857	1.220	1.075	0.778	1.382	14.530	10.405	1.396	550
600	1.075	0.886	1.213	1.087	0.790	1.376	14.546	10.422	1.396	600
650	1.102	0.913	1.207	1.100	0.803	1.370	14.571	10.447	1.395	650
700	1.126	0.937	1.202	1.113	0.816	1.364	14.604	10.480	1.394	700
750	1.148	0.959	1.197	1.126	0.829	1.358	14.645	10.521	1.392	750
800	1.169	0.980	1.193	1.139	0.842	1.353	14.695	10.570	1.390	800
900	1.204	1.015	1.186	1.163	0.866	1.343	14.822	10.698	1.385	900
1000	1.234	1.045	1.181	1.185	0.888	1.335	14.983	10.859	1.380	1000

Source: Adapted from K. Wark, *Thermodynamics, 4th ed.,* McGraw-Hill, New York, 1983, as based on "Tables of Thermal Properties of Gases," NBS Circular 564, 1955.

TABLE A-15 Variation of $\bar{c}_p$ with Temperature for Various Ideal Gases

$$\frac{\bar{c}_p}{\bar{R}} = \alpha + \beta T + \gamma T^2 + \delta T^3 + \varepsilon T^4$$

T is in K, equations valid from 300 to 1000 K

Gas	α	$\beta \times 10^3$	$\gamma \times 10^6$	$\delta \times 10^9$	$\epsilon \times 10^{12}$
CO	3.710	−1.619	3.692	−2.032	0.240
CO_2	2.401	8.735	−6.607	2.002	0
H_2	3.057	2.677	−5.810	5.521	−1.812
H_2O	4.070	−1.108	4.152	−2.964	0.807
O_2	3.626	−1.878	7.055	−6.764	2.156
N_2	3.675	−1.208	2.324	−0.632	−0.226
Air	3.653	−1.337	3.294	−1.913	0.2763
SO_2	3.267	5.324	0.684	−5.281	2.559
CH_4	3.826	−3.979	24.558	−22.733	6.963
C_2H_2	1.410	19.057	−24.501	16.391	−4.135
C_2H_4	1.426	11.383	7.989	−16.254	6.749
Monatomic gases[a]	2.5	0	0	0	0

[a] For monatomic gases, such as He, Ne, and Ar, $\bar{c}_p$ is constant over a wide temperature range and is very nearly equal to $5/2\,\bar{R}$.

Source: Adapted from K. Wark, *Thermodynamics, 4th ed.,* McGraw-Hill, New York, 1983, as based on NASA SP-273, U.S. Government Printing Office, Washington, DC, 1971.

TABLE A-16 Ideal Gas Properties of Air

T	h	p_r	u	v_r	s^o	T	h	p_r	u	v_r	s^o
200	199.97	0.3363	142.56	1707.	1.29559	450	451.80	5.775	322.62	223.6	2.11161
210	209.97	0.3987	149.69	1512.	1.34444	460	462.02	6.245	329.97	211.4	2.13407
220	219.97	0.4690	156.82	1346.	1.39105	470	472.24	6.742	337.32	200.1	2.15604
230	230.02	0.5477	164.00	1205.	1.43557	480	482.49	7.268	344.70	189.5	2.17760
240	240.02	0.6355	171.13	1084.	1.47824	490	492.74	7.824	352.08	179.7	2.19876
250	250.05	0.7329	178.28	979.	1.51917	500	503.02	8.411	359.49	170.6	2.21952
260	260.09	0.8405	185.45	887.8	1.55848	510	513.32	9.031	366.92	162.1	2.23993
270	270.11	0.9590	192.60	808.0	1.59634	520	523.63	9.684	374.36	154.1	2.25997
280	280.13	1.0889	199.75	738.0	1.63279	530	533.98	10.37	381.84	146.7	2.27967
285	285.14	1.1584	203.33	706.1	1.65055	540	544.35	11.10	389.34	139.7	2.29906
290	290.16	1.2311	206.91	676.1	1.66802	550	554.74	11.86	396.86	133.1	2.31809
295	295.17	1.3068	210.49	647.9	1.68515	560	565.17	12.66	404.42	127.0	2.33685
300	300.19	1.3860	214.07	621.2	1.70203	570	575.59	13.50	411.97	121.2	2.35531
305	305.22	1.4686	217.67	596.0	1.71865	580	586.04	14.38	419.55	115.7	2.37348
310	310.24	1.5546	221.25	572.3	1.73498	590	596.52	15.31	427.15	110.6	2.39140
315	315.27	1.6442	224.85	549.8	1.75106	600	607.02	16.28	434.78	105.8	2.40902
320	320.29	1.7375	228.42	528.6	1.76690	610	617.53	17.30	442.42	101.2	2.42644
325	325.31	1.8345	232.02	508.4	1.78249	620	628.07	18.36	450.09	96.92	2.44356
330	330.34	1.9352	235.61	489.4	1.79783	630	638.63	19.84	457.78	92.84	2.46048
340	340.42	2.149	242.82	454.1	1.82790	640	649.22	20.64	465.50	88.99	2.47716
350	350.49	2.379	250.02	422.2	1.85708	650	659.84	21.86	473.25	85.34	2.49364
360	360.58	2.626	257.24	393.4	1.88543	660	670.47	23.13	481.01	81.89	2.50985
370	370.67	2.892	264.46	367.2	1.91313	670	681.14	24.46	488.81	78.61	2.52589
380	380.77	3.176	271.69	343.4	1.94001	680	691.82	25.85	496.62	75.50	2.54175
390	390.88	3.481	278.93	321.5	1.96633	690	702.52	27.29	504.45	72.56	2.55731
400	400.98	3.806	286.16	301.6	1.99194	700	713.27	28.80	512.33	69.76	2.57277
410	411.12	4.153	293.43	283.3	2.01699	710	724.04	30.38	520.23	67.07	2.58810
420	421.26	4.522	300.69	266.6	2.04142	720	734.82	32.02	528.14	64.53	2.60319
430	431.43	4.915	307.99	251.1	2.06533	730	745.62	33.72	536.07	62.13	2.61803
440	441.61	5.332	315.30	236.8	2.08870	740	756.44	35.50	544.02	59.82	2.63280

TABLE A-16 (Continued)

T (K), h and u (kJ/kg), s^o (kJ/kg·K)

T	h	p_r	u	v_r	s^o
750	767.29	37.35	551.99	57.63	2.64737
760	778.18	39.27	560.01	55.54	2.66176
770	789.11	41.31	568.07	53.39	2.67595
780	800.03	43.35	576.12	51.64	2.69013
790	810.99	45.55	584.21	49.86	2.70400
800	821.95	47.75	592.30	48.08	2.71787
820	843.98	52.59	608.59	44.84	2.74504
840	866.08	57.60	624.95	41.85	2.77170
860	888.27	63.09	641.40	39.12	2.79783
880	910.56	68.98	657.95	36.61	2.82344
900	932.93	75.29	674.58	34.31	2.84856
920	955.38	82.05	691.28	32.18	2.87324
940	977.92	89.28	708.08	30.22	2.89748
960	1000.55	97.00	725.02	28.40	2.92128
980	1023.25	105.2	741.98	26.73	2.94468
1000	1046.04	114.0	758.94	25.17	2.96770
1020	1068.89	123.4	776.10	23.72	2.99034
1040	1091.85	133.3	793.36	22.39	3.01260
1060	1114.86	143.9	810.62	21.14	3.03449
1080	1137.89	155.2	827.88	19.98	3.05608
1100	1161.07	167.1	845.33	18.896	3.07732
1120	1184.28	179.7	862.79	17.886	3.09825
1140	1207.57	193.1	880.35	16.946	3.11883
1160	1230.92	207.2	897.91	16.064	3.13916
1180	1254.34	222.2	915.57	15.241	3.15916
1200	1277.79	238.0	933.33	14.470	3.17888
1220	1301.31	254.7	951.09	13.747	3.19834
1240	1324.93	272.3	968.95	13.069	3.21751
1260	1348.55	290.8	986.90	12.435	3.23638
1280	1372.24	310.4	1004.76	11.835	3.25510
1300	1395.97	330.9	1022.82	11.275	3.27345
1320	1419.76	352.5	1040.88	10.747	3.29160
1340	1443.60	375.3	1058.94	10.247	3.30959
1360	1467.49	399.1	1077.10	9.780	3.32724
1380	1491.44	424.2	1095.26	9.337	3.34474
1400	1515.42	450.5	1113.52	8.919	3.36200
1420	1539.44	478.0	1131.77	8.526	3.37901
1440	1563.51	506.9	1150.13	8.153	3.39586
1460	1587.63	537.1	1168.49	7.801	3.41247
1480	1611.79	568.8	1186.95	7.468	3.42892
1500	1635.97	601.9	1205.41	7.152	3.44516
1520	1660.23	636.5	1223.87	6.854	3.46120
1540	1684.51	672.8	1242.43	6.569	3.47712
1560	1708.82	710.5	1260.99	6.301	3.49276
1580	1733.17	750.0	1279.65	6.046	3.50829
1600	1757.57	791.2	1298.30	5.804	3.52364
1620	1782.00	834.1	1316.96	5.574	3.53879
1640	1806.46	878.9	1335.72	5.355	3.55381
1660	1830.96	925.6	1354.48	5.147	3.56867
1680	1855.50	974.2	1373.24	4.949	3.58335
1700	1880.1	1025	1392.7	4.761	3.5979
1750	1941.6	1161	1439.8	4.328	3.6336
1800	2003.3	1310	1487.2	3.944	3.6684
1850	2065.3	1475	1534.9	3.601	3.7023
1900	2127.4	1655	1582.6	3.295	3.7354
1950	2189.7	1852	1630.6	3.022	3.7677
2000	2252.1	2068	1678.7	2.776	3.7994
2050	2314.6	2303	1726.8	2.555	3.8303
2100	2377.4	2559	1775.3	2.356	3.8605
2150	2440.3	2837	1823.8	2.175	3.8901
2200	2503.2	3138	1872.4	2.012	3.9191
2250	2566.4	3464	1921.3	1.864	3.9474

Source: Adapted from K. Wark, *Thermodynamics, 4th ed.*, McGraw-Hill, New York, 1983, as based on J. H. Keenan and J. Kaye, "Gas Tables," Wiley, New York, 1945.

TABLE A-17 Ideal Gas Properties of Nitrogen, N_2

$T(K)$, $\bar{h}$ and $\bar{u}(kJ/kmol)$, $\bar{s}°(kJ/kmol \cdot K)$

$[\bar{h}_f° = 0 \ kJ/kmol]$

T	$\bar{h}$	$\bar{u}$	$\bar{s}°$	T	$\bar{h}$	$\bar{u}$	$\bar{s}°$
0	0	0	0	600	17,563	12,574	212.066
220	6,391	4,562	182.639	610	17,864	12,792	212.564
230	6,683	4,770	183.938	620	18,166	13,011	213.055
240	6,975	4,979	185.180	630	18,468	13,230	213.541
250	7,266	5,188	186.370	640	18,772	13,450	214.018
260	7,558	5,396	187.514	650	19,075	13,671	214.489
270	7,849	5,604	188.614	660	19,380	13,892	214.954
280	8,141	5,813	189.673	670	19,685	14,114	215.413
290	8,432	6,021	190.695	680	19,991	14,337	215.866
298	8,669	6,190	191.502	690	20,297	14,560	216.314
300	8,723	6,229	191.682	700	20,604	14,784	216.756
310	9,014	6,437	192.638	710	20,912	15,008	217.192
320	9,306	6,645	193.562	720	21,220	15,234	217.624
330	9,597	6,853	194.459	730	21,529	15,460	218.059
340	9,888	7,061	195.328	740	21,839	15,686	218.472
350	10,180	7,270	196.173	750	22,149	15,913	218.889
360	10,471	7,478	196.995	760	22,460	16,141	219.301
370	10,763	7,687	197.794	770	22,772	16,370	219.709
380	11,055	7,895	198.572	780	23,085	16,599	220.113
390	11,347	8,104	199.331	790	23,398	16,830	220.512
400	11,640	8,314	200.071	800	23,714	17,061	220.907
410	11,932	8,523	200.794	810	24,027	17,292	221.298
420	12,225	8,733	201.499	820	24,342	17,524	221.684
430	12,518	8,943	202.189	830	24,658	17,757	222.067
440	12,811	9,153	202.863	840	24,974	17,990	222.447
450	13,105	9,363	203.523	850	25,292	18,224	222.822
460	13,399	9,574	204.170	860	25,610	18,459	223.194
470	13,693	9,786	204.803	870	25,928	18,695	223.562
480	13,988	9,997	205.424	880	26,248	18,931	223.927
490	14,285	10,210	206.033	890	26,568	19,168	224.288
500	14,581	10,423	206.630	900	26,890	19,407	224.647
510	14,876	10,635	207.216	910	27,210	19,644	225.002
520	15,172	10,848	207.792	920	27,532	19,883	225.353
530	15,469	11,062	208.358	930	27,854	20,122	225.701
540	15,766	11,277	208.914	940	28,178	20,362	226.047
550	16,064	11,492	209.461	950	28,501	20,603	226.389
560	16,363	11,707	209.999	960	28,826	20,844	226.728
570	16,662	11,923	210.528	970	29,151	21,086	227.064
580	16,962	12,139	211.049	980	29,476	21,328	227.398
590	17,262	12,356	211.562	990	29,803	21,571	227.728

TABLE A-17 (*Continued*)

T	$\bar{h}$	$\bar{u}$	$\bar{s}°$	T	$\bar{h}$	$\bar{u}$	$\bar{s}°$
1000	30,129	21,815	228.057	1760	56,227	41,594	247.396
1020	30,784	22,304	228.706	1780	56,938	42,139	247.798
1040	31,442	22,795	229.344	1800	57,651	42,685	248.195
1060	32,101	23,288	229.973	1820	58,363	43,231	248.589
1080	32,762	23,782	230.591	1840	59,075	43,777	248.979
1100	33,426	24,280	231.199	1860	59,790	44,324	249.365
1120	34,092	24,780	231.799	1880	60,504	44,873	249.748
1140	34,760	25,282	232.391	1900	61,220	45,423	250.128
1160	35,430	25,786	232.973	1920	61,936	45,973	250.502
1180	36,104	26,291	233.549	1940	62,654	46,524	250.874
1200	36,777	26,799	234.115	1960	63,381	47,075	251.242
1220	37,452	27,308	234.673	1980	64,090	47,627	251.607
1240	38,129	27,819	235.223	2000	64,810	48,181	251.969
1260	38,807	28,331	235.766	2050	66,612	49,567	252.858
1280	39,488	28,845	236.302	2100	68,417	50,957	253.726
1300	40,170	29,361	236.831	2150	70,226	52,351	254.578
1320	40,853	29,878	237.353	2200	72,040	53,749	255.412
1340	41,539	30,398	237.867	2250	73,856	55,149	256.227
1360	42,227	30,919	238.376	2300	75,676	56,553	257.027
1380	42,915	31,441	238.878	2350	77,496	57,958	257.810
1400	43,605	31,964	239.375	2400	79,320	59,366	258.580
1420	44,295	32,489	239.865	2450	81,149	60,779	259.332
1440	44,988	33,014	240.350	2500	82,981	62,195	260.073
1460	45,682	33,543	240.827	2550	84,814	63,613	260.799
1480	46,377	34,071	241.301	2600	86,650	65,033	261.512
1500	47,073	34,601	241.768	2650	88,488	66,455	262.213
1520	47,771	35,133	242.228	2700	90,328	67,880	262.902
1540	48,470	35,665	242.685	2750	92,171	69,306	263.577
1560	49,168	36,197	243.137	2800	94,014	70,734	264.241
1580	49,869	36,732	243.585	2850	95,859	72,163	264.895
1600	50,571	37,268	244.028	2900	97,705	73,593	265.538
1620	51,275	37,806	244.464	2950	99,556	75,028	266.170
1640	51,980	38,344	244.896	3000	101,407	76,464	266.793
1660	52,686	38,884	245.324	3050	103,260	77,902	267.404
1680	53,393	39,424	245.747	3100	105,115	79,341	268.007
1700	54,099	39,965	246.166	3150	106,972	80,782	268.601
1720	54,807	40,507	246.580	3200	108,830	82,224	269.186
1740	55,516	41,049	246.990	3250	110,690	83,668	269.763

Source: Tables A-14 through A-19 are adapted from K. Wark, *Thermodynamics, 4th ed.*, McGraw-Hill, New York, 1983, as based on the JANAF Thermochemical Tables, NSRDS-NBS-37, 1971.

TABLE A-18 Ideal Gas Properties of Oxygen, O_2

T(K), $\bar{h}$ and $\bar{u}$(kJ/kmol), $\bar{s}°$(kJ/kmol · K)

$[\bar{h}_f° = 0$ kJ/kmol$]$

T	$\bar{h}$	$\bar{u}$	$\bar{s}°$	T	$\bar{h}$	$\bar{u}$	$\bar{s}°$
0	0	0	0	600	17,929	12,940	226.346
220	6,404	4,575	196.171	610	18,250	13,178	226.877
230	6,694	4,782	197.461	620	18,572	13,417	227.400
240	6,984	4,989	198.696	630	18,895	13,657	227.918
250	7,275	5,197	199.885	640	19,219	13,898	228.429
260	7,566	5,405	201.027	650	19,544	14,140	228.932
270	7,858	5,613	202.128	660	19,870	14,383	229.430
280	8,150	5,822	203.191	670	20,197	14,626	229.920
290	8,443	6,032	204.218	680	20,524	14,871	230.405
298	8,682	6,203	205.033	690	20,854	15,116	230.885
300	8,736	6,242	205.213	700	21,184	15,364	231.358
310	9,030	6,453	206.177	710	21,514	15,611	231.827
320	9,325	6,664	207.112	720	21,845	15,859	232.291
330	9,620	6,877	208.020	730	22,177	16,107	232.748
340	9,916	7,090	208.904	740	22,510	16,357	233.201
350	10,213	7,303	209.765	750	22,844	16,607	233.649
360	10,511	7,518	210.604	760	23,178	16,859	234.091
370	10,809	7,733	211.423	770	23,513	17,111	234.528
380	11,109	7,949	212.222	780	23,850	17,364	234.960
390	11,409	8,166	213.002	790	24,186	17,618	235.387
400	11,711	8,384	213.765	800	24,523	17,872	235.810
410	12,012	8,603	214.510	810	24,861	18,126	236.230
420	12,314	8,822	215.241	820	25,199	18,382	236.644
430	12,618	9,043	215.955	830	25,537	18,637	237.055
440	12,923	9,264	216.656	840	25,877	18,893	237.462
450	13,228	9,487	217.342	850	26,218	19,150	237.864
460	13,535	9,710	218.016	860	26,559	19,408	238.264
470	13,842	9,935	218.676	870	26,899	19,666	238.660
480	14,151	10,160	219.326	880	27,242	19,925	239.051
490	14,460	10,386	219.963	890	27,584	20,185	239.439
500	14,770	10,614	220.589	900	27,928	20,445	239.823
510	15,082	10,842	221.206	910	28,272	20,706	240.203
520	15,395	11,071	221.812	920	28,616	20,967	240.580
530	15,708	11,301	222.409	930	28,960	21,228	240.953
540	16,022	11,533	222.997	940	29,306	21,491	241.323
550	16,338	11,765	223.576	950	29,652	21,754	241.689
560	16,654	11,998	224.146	960	29,999	22,017	242.052
570	16,971	12,232	224.708	970	30,345	22,280	242.411
580	17,290	12,467	225.262	980	30,692	22,544	242.768
590	17,609	12,703	225.808	990	31,041	22,809	243.120

TABLE A-18 (*Continued*)

T	$\bar{h}$	$\bar{u}$	$\bar{s}°$	T	$\bar{h}$	$\bar{u}$	$\bar{s}°$
1000	31,389	23,075	243.471	1760	58,880	44,247	263.861
1020	32,088	23,607	244.164	1780	59,624	44,825	264.283
1040	32,789	24,142	244.844	1800	60,371	45,405	264.701
1060	33,490	24,677	245.513	1820	61,118	45,986	265.113
1080	34,194	25,214	246.171	1840	61,866	46,568	265.521
1100	34,899	25,753	246.818	1860	62,616	47,151	265.925
1120	35,606	26,294	247.454	1880	63,365	47,734	266.326
1140	36,314	26,836	248.081	1900	64,116	48,319	266.722
1160	37,023	27,379	248.698	1920	64,868	48,904	267.115
1180	37,734	27,923	249.307	1940	65,620	49,490	267.505
1200	38,447	28,469	249.906	1960	66,374	50,078	267.891
1220	39,162	29,018	250.497	1980	67,127	50,665	268.275
1240	39,877	29,568	251.079	2000	67,881	51,253	268.655
1260	40,594	30,118	251.653	2050	69,772	52,727	269.588
1280	41,312	30,670	252.219	2100	71,668	54,208	270.504
1300	42,033	31,224	252.776	2150	73,573	55,697	271.399
1320	42,753	31,778	253.325	2200	75,484	57,192	272.278
1340	43,475	32,334	253.868	2250	77,397	58,690	273.136
1360	44,198	32,891	254.404	2300	79,316	60,193	273.981
1380	44,923	33,449	254.932	2350	81,243	61,704	274.809
1400	45,648	34,008	255.454	2400	83,174	63,219	275.625
1420	46,374	34,567	255.968	2450	85,112	64,742	276.424
1440	47,102	35,129	256.475	2500	87,057	66,271	277.207
1460	47,831	35,692	256.978	2550	89,004	67,802	277.979
1480	48,561	36,256	257.474	2600	90,956	69,339	278.738
1500	49,292	36,821	257.965	2650	92,916	70,883	279.485
1520	50,024	37,387	258.450	2700	94,881	72,433	280.219
1540	50,756	37,952	258.928	2750	96,852	73,987	280.942
1560	51,490	38,520	259.402	2800	98,826	75,546	281.654
1580	52,224	39,088	259.870	2850	100,808	77,112	282.357
1600	52,961	39,658	260.333	2900	102,793	78,682	283.048
1620	53,696	40,227	260.791	2950	104,785	80,258	283.728
1640	54,434	40,799	261.242	3000	106,780	81,837	284.399
1660	55,172	41,370	261.690	3050	108,778	83,419	285.060
1680	55,912	41,944	262.132	3100	110,784	85,009	285.713
1700	56,652	42,517	262.571	3150	112,795	86,601	286.355
1720	57,394	43,093	263.005	3200	114,809	88,203	286.989
1740	58,136	43,669	263.435	3250	116,827	89,804	287.614

TABLE A-19 Ideal Gas Properties of Water Vapor, H_2O

$T(K)$, $\bar{h}$ and $\bar{u}(kJ/kmol)$, $\bar{s}°(kJ/kmol \cdot K)$

$[\bar{h}_f° = -241,820 \text{ kJ/kmol}]$

T	$\bar{h}$	$\bar{u}$	$\bar{s}°$	T	$\bar{h}$	$\bar{u}$	$\bar{s}°$
0	0	0	0	600	20,402	15,413	212.920
220	7,295	5,466	178.576	610	20,765	15,693	213.529
230	7,628	5,715	180.054	620	21,130	15,975	214.122
240	7,961	5,965	181.471	630	21,495	16,257	214.707
250	8,294	6,215	182.831	640	21,862	16,541	215.285
260	8,627	6,466	184.139	650	22,230	16,826	215.856
270	8,961	6,716	185.399	660	22,600	17,112	216.419
280	9,296	6,968	186.616	670	22,970	17,399	216.976
290	9,631	7,219	187.791	680	23,342	17,688	217.527
298	9,904	7,425	188.720	690	23,714	17,978	218.071
300	9,966	7,472	188.928	700	24,088	18,268	218.610
310	10,302	7,725	190.030	710	24,464	18,561	219.142
320	10,639	7,978	191.098	720	24,840	18,854	219.668
330	10,976	8,232	192.136	730	25,218	19,148	220.189
340	11,314	8,487	193.144	740	25,597	19,444	220.707
350	11,652	8,742	194.125	750	25,977	19,741	221.215
360	11,992	8,998	195.081	760	26,358	20,039	221.720
370	12,331	9,255	196.012	770	26,741	20,339	222.221
380	12,672	9,513	196.920	780	27,125	20,639	222.717
390	13,014	9,771	197.807	790	27,510	20,941	223.207
400	13,356	10,030	198.673	800	27,896	21,245	223.693
410	13,699	10,290	199.521	810	28,284	21,549	224.174
420	14,043	10,551	200.350	820	28,672	21,855	224.651
430	14,388	10,813	201.160	830	29,062	22,162	225.123
440	14,734	11,075	201.955	840	29,454	22,470	225.592
450	15,080	11,339	202.734	850	29,846	22,779	226.057
460	15,428	11,603	203.497	860	30,240	23,090	226.517
470	15,777	11,869	204.247	870	30,635	23,402	226.973
480	16,126	12,135	204.982	880	31,032	23,715	227.426
490	16,477	12,403	205.705	890	31,429	24,029	227.875
500	16,828	12,671	206.413	900	31,828	24,345	228.321
510	17,181	12,940	207.112	910	32,228	24,662	228.763
520	17,534	13,211	207.799	920	32,629	24,980	229.202
530	17,889	13,482	208.475	930	33,032	25,300	229.637
540	18,245	13,755	209.139	940	33,436	25,621	230.070
550	18,601	14,028	209.795	950	33,841	25,943	230.499
560	18,959	14,303	210.440	960	34,247	26,265	230.924
570	19,318	14,579	211.075	970	34,653	26,588	231.347
580	19,678	14,856	211.702	980	35,061	26,913	231.767
590	20,039	15,134	212.320	990	35,472	27,240	232.184

TABLE A-19 (*Continued*)

T	$\bar{h}$	$\bar{u}$	$\bar{s}°$	T	$\bar{h}$	$\bar{u}$	$\bar{s}°$
1000	35,882	27,568	232.597	1760	70,535	55,902	258.151
1020	36,709	28,228	233.415	1780	71,523	56,723	258.708
1040	37,542	28,895	234.223	1800	72,513	57,547	259.262
1060	38,380	29,567	235.020	1820	73,507	58,375	259.811
1080	39,223	30,243	235.806	1840	74,506	59,207	260.357
1100	40,071	30,925	236.584	1860	75,506	60,042	260.898
1120	40,923	31,611	237.352	1880	76,511	60,880	261.436
1140	41,780	32,301	238.110	1900	77,517	61,720	261.969
1160	42,642	32,997	238.859	1920	78,527	62,564	262.497
1180	43,509	33,698	239.600	1940	79,540	63,411	263.022
1200	44,380	34,403	240.333	1960	80,555	64,259	263.542
1220	45,256	35,112	241.057	1980	81,573	65,111	264.059
1240	46,137	35,827	241.773	2000	82,593	65,965	264.571
1260	47,022	36,546	242.482	2050	85,156	68,111	265.838
1280	47,912	37,270	243.183	2100	87,735	70,275	267.081
1300	48,807	38,000	243.877	2150	90,330	72,454	268.301
1320	49,707	38,732	244.564	2200	92,940	74,649	269.500
1340	50,612	39,470	245.243	2250	95,562	76,855	270.679
1360	51,521	40,213	245.915	2300	98,199	79,076	271.839
1380	52,434	40,960	246.582	2350	100,846	81,308	272.978
1400	53,351	41,711	247.241	2400	103,508	83,553	274.098
1420	54,273	42,466	247.895	2450	106,183	85,811	275.201
1440	55,198	43,226	248.543	2500	108,868	88,082	276.286
1460	56,128	43,989	249.185	2550	111,565	90,364	277.354
1480	57,062	44,756	249.820	2600	114,273	92,656	278.407
1500	57,999	45,528	250.450	2650	116,991	94,958	279.441
1520	58,942	46,304	251.074	2700	119,717	97,269	280.462
1540	59,888	47,084	251.693	2750	122,453	99,588	281.464
1560	60,838	47,868	252.305	2800	125,198	101,917	282.453
1580	61,792	48,655	252.912	2850	127,952	104,256	283.429
1600	62,748	49,445	253.513	2900	130,717	106,605	284.390
1620	63,709	50,240	254.111	2950	133,486	108,959	285.338
1640	64,675	51,039	254.703	3000	136,264	111,321	286.273
1660	65,643	51,841	255.290	3050	139,051	113,692	287.194
1680	66,614	52,646	255.873	3100	141,846	116,072	288.102
1700	67,589	53,455	256.450	3150	144,648	118,458	288.999
1720	68,567	54,267	257.022	3200	147,457	120,851	289.884
1740	69,550	55,083	257.589	3250	150,272	123,250	290.756

TABLE A-20 Ideal Gas Properties of Carbon Monoxide, CO

$T(K)$, $\bar{h}$ and $\bar{u}(kJ/kmol)$, $\bar{s}°(kJ/kmol \cdot K)$

$[\bar{h}_f° = -110{,}530 \text{ kJ/kmol}]$

T	$\bar{h}$	$\bar{u}$	$\bar{s}°$	T	$\bar{h}$	$\bar{u}$	$\bar{s}°$
0	0	0	0	600	17,611	12,622	218.204
220	6,391	4,562	188.683	610	17,915	12,843	218.708
230	6,683	4,771	189.980	620	18,221	13,066	219.205
240	6,975	4,979	191.221	630	18,527	13,289	219.695
250	7,266	5,188	192.411	640	18,833	13,512	220.179
260	7,558	5,396	193.554	650	19,141	13,736	220.656
270	7,849	5,604	194.654	660	19,449	13,962	221.127
280	8,140	5,812	195.173	670	19,758	14,187	221.592
290	8,432	6,020	196.735	680	20,068	14,414	222.052
298	8,669	6,190	197.543	690	20,378	14,641	222.505
300	8,723	6,229	197.723	700	20,690	14,870	222.953
310	9,014	6,437	198.678	710	21,002	15,099	223.396
320	9,306	6,645	199.603	720	21,315	15,328	223.833
330	9,597	6,854	200.500	730	21,628	15,558	224.265
340	9,889	7,062	201.371	740	21,943	15,789	224.692
350	10,181	7,271	202.217	750	22,258	16,022	225.115
360	10,473	7,480	203.040	760	22,573	16,255	225.533
370	10,765	7,689	203.842	770	22,890	16,488	225.947
380	11,058	7,899	204.622	780	23,208	16,723	226.357
390	11,351	8,108	205.383	790	23,526	16,957	226.762
400	11,644	8,319	206.125	800	23,844	17,193	227.162
410	11,938	8,529	206.850	810	24,164	17,429	227.559
420	12,232	8,740	207.549	820	24,483	17,665	227.952
430	12,526	8,951	208.252	830	24,803	17,902	228.339
440	12,821	9,163	208.929	840	25,124	18,140	228.724
450	13,116	9,375	209.593	850	25,446	18,379	229.106
460	13,412	9,587	210.243	860	25,768	18,617	229.482
470	13,708	9,800	210.880	870	26,091	18,858	229.856
480	14,005	10,014	211.504	880	26,415	19,099	230.227
490	14,302	10,228	212.117	890	26,740	19,341	230.593
500	14,600	10,443	212.719	900	27,066	19,583	230.957
510	14,898	10,658	213.310	910	27,392	19,826	231.317
520	15,197	10,874	213.890	920	27,719	20,070	231.674
530	15,497	11,090	214.460	930	28,046	20,314	232.028
540	15,797	11,307	215.020	940	28,375	20,559	232.379
550	16,097	11,524	215.572	950	28,703	20,805	232.727
560	16,399	11,743	216.115	960	29,033	21,051	233.072
570	16,701	11,961	216.649	970	29,362	21,298	233.413
580	17,003	12,181	217.175	980	29,693	21,545	233.752
590	17,307	12,401	217.693	990	30,024	21,793	234.088

TABLE A-20 (*Continued*)

T	$\bar{h}$	$\bar{u}$	$\bar{s}°$	T	$\bar{h}$	$\bar{u}$	$\bar{s}°$
1000	30,355	22,041	234.421	1760	56,756	42,123	253.991
1020	31,020	22,540	235.079	1780	57,473	42,673	254.398
1040	31,688	23,041	235.728	1800	58,191	43,225	254.797
1060	32,357	23,544	236.364	1820	58,910	43,778	255.194
1080	33,029	24,049	236.992	1840	59,629	44,331	255.587
1100	33,702	24,557	237.609	1860	60,351	44,886	255.976
1120	34,377	25,065	238.217	1880	61,072	45,441	256.361
1140	35,054	25,575	238.817	1900	61,794	45,997	256.743
1160	35,733	26,088	239.407	1920	62,516	46,552	257.122
1180	36,406	26,602	239.989	1940	63,238	47,108	257.497
1200	37,095	27,118	240.663	1960	63,961	47,665	257.868
1220	37,780	27,637	241.128	1980	64,684	48,221	258.236
1240	38,466	28,426	241.686	2000	65,408	48,780	258.600
1260	39,154	28,678	242.236	2050	67,224	50,179	259.494
1280	39,884	29,201	242.780	2100	69,044	51,584	260.370
1300	40,534	29,725	243.316	2150	70,864	52,988	261.226
1320	41,266	30,251	243.844	2200	72,688	54,396	262.065
1340	41,919	30,778	244.366	2250	74,516	55,809	262.887
1360	42,613	31,306	244.880	2300	76,345	57,222	263.692
1380	43,309	31,836	245.388	2350	78,178	58,640	264.480
1400	44,007	32,367	**245.889**	2400	80,015	60,060	265.253
1420	44,707	32,900	**246.385**	2450	81,852	61,482	266.012
1440	45,408	33,434	246.876	2500	83,692	62,906	266.755
1460	46,110	33,971	247.360	2550	85,537	64,335	267.485
1480	46,813	34,508	247.839	2600	87,383	65,766	268.202
1500	47,517	35,046	248.312	2650	89,230	67,197	268.905
1520	48,222	35,584	248.778	2700	91,077	68,628	269.596
1540	48,928	36,124	249.240	2750	92,930	70,066	270.285
1560	49,635	36,665	249.695	2800	94,784	71,504	270.943
1580	50,344	37,207	250.147	2850	96,639	72,945	271.602
1600	51,053	37,750	250.592	2900	98,495	74,383	272.249
1620	51,763	38,293	251.033	2950	100,352	75,825	272.884
1640	52,472	38,837	251.470	3000	102,210	77,267	273.508
1660	53,184	39,382	251.901	3050	104,073	78,715	274.123
1680	53,895	39,927	252.329	3100	105,939	80,164	274.730
1700	54,609	40,474	252.751	3150	107,802	81,612	275.326
1720	55,323	41,023	253.169	3200	109,667	83,061	275.914
1740	56,039	41,572	253.582	3250	111,534	84,513	276.494

TABLE A-21 Ideal Gas Properties of Carbon Dioxide, CO_2

$T(K)$, $\bar{h}$ and $\bar{u}(kJ/kmol)$, $\bar{s}°(kJ/kmol \cdot K)$

$[\bar{h}_f° = -393,520 \; kJ/kmol]$

T	$\bar{h}$	$\bar{u}$	$\bar{s}°$	T	$\bar{h}$	$\bar{u}$	$\bar{s}°$
0	0	0	0	600	22,280	17,291	243.199
220	6,601	4,772	202.966	610	22,754	17,683	243.983
230	6,938	5,026	204.464	620	23,231	18,076	244.758
240	7,280	5,285	205.920	630	23,709	18,471	245.524
250	7,627	5,548	207.337	640	24,190	18,869	246.282
260	7,979	5,817	208.717	650	24,674	19,270	247.032
270	8,335	6,091	210.062	660	25,160	19,672	247.773
280	8,697	6,369	211.376	670	25,648	20,078	248.507
290	9,063	6,651	212.660	680	26,138	20,484	249.233
298	9,364	6,885	213.685	690	26,631	20,894	249.952
300	9,431	6,939	213.915	700	27,125	21,305	250.663
310	9,807	7,230	215.146	710	27,622	21,719	251.368
320	10,186	7,526	216.351	720	28,121	22,134	252.065
330	10,570	7,826	217.534	730	28,622	22,552	252.755
340	10,959	8,131	218.694	740	29,124	22,972	253.439
350	11,351	8,439	219.831	750	29,629	23,393	254.117
360	11,748	8,752	220.948	760	30,135	23,817	254.787
370	12,148	9,068	222.044	770	30,644	24,242	255.452
380	12,552	9,392	223.122	780	31,154	24,669	256.110
390	12,960	9,718	224.182	790	31,665	25,097	256.762
400	13,372	10,046	225.225	800	32,179	25,527	257.408
410	13,787	10,378	226.250	810	32,694	25,959	258.048
420	14,206	10,714	227.258	820	33,212	26,394	258.682
430	14,628	11,053	228.252	830	33,730	26,829	259.311
440	15,054	11,393	229.230	840	34,251	27,267	259.934
450	15,483	11,742	230.194	850	34,773	27,706	260.551
460	15,916	12,091	231.144	860	35,296	28,125	261.164
470	16,351	12,444	232.080	870	35,821	28,588	261.770
480	16,791	12,800	233.004	880	36,347	29,031	262.371
490	17,232	13,158	233.916	890	36,876	29,476	262.968
500	17,678	13,521	234.814	900	37,405	29,922	263.559
510	18,126	13,885	235.700	910	37,935	30,369	264.146
520	18,576	14,253	236.575	920	38,467	30,818	264.728
530	19,029	14,622	237.439	930	39,000	31,268	265.304
540	19,485	14,996	238.292	940	39,535	31,719	265.877
550	19,945	15,372	239.135	950	40,070	32,171	266.444
560	20,407	15,751	239.962	960	40,607	32,625	267.007
570	20,870	16,131	240.789	970	41,145	33,081	267.566
580	21,337	16,515	241.602	980	41,685	33,537	268.119
590	21,807	16,902	242.405	990	42,226	33,995	268.670

TABLE A-21 (*Continued*)

T	$\bar{h}$	$\bar{u}$	$\bar{s}^\circ$	T	$\bar{h}$	$\bar{u}$	$\bar{s}^\circ$
1000	42,769	34,455	269.215	1760	86,420	71,787	301.543
1020	43,859	35,378	270.293	1780	87,612	72,812	302.271
1040	44,953	36,306	271.354	1800	88,806	73,840	302.884
1060	46,051	37,238	272.400	1820	90,000	74,868	303.544
1080	47,153	38,174	273.430	1840	91,196	75,897	304.198
1100	48,258	39,112	274.445	1860	92,394	76,929	304.845
1120	49,369	40,057	275.444	1880	93,593	77,962	305.487
1140	50,484	41,006	276.430	1900	94,793	78,996	306.122
1160	51,602	41,957	277.403	1920	95,995	80,031	306.751
1180	52,724	42,913	278.362	1940	97,197	81,067	307.374
1200	53,848	43,871	279.307	1960	98,401	82,105	307.992
1220	54,977	44,834	280.238	1980	99,606	83,144	308.604
1240	56,108	45,799	281.158	2000	100,804	84,185	309.210
1260	57,244	46,768	282.066	2050	103,835	86,791	310.701
1280	58,381	47,739	282.962	2100	106,864	89,404	312.160
1300	59,522	48,713	283.847	2150	109,898	92,023	313.589
1320	60,666	49,691	284.722	2200	112,939	94,648	314.988
1340	61,813	50,672	285.586	2250	115,984	97,277	316.356
1360	62,963	51,656	286.439	2300	119,035	99,912	317.695
1380	64,116	52,643	287.283	2350	122,091	102,552	319.011
1400	65,271	53,631	288.106	2400	125,152	105,197	320.302
1420	66,427	54,621	288.934	2450	128,219	107,849	321.566
1440	67,586	55,614	289.743	2500	131,290	110,504	322.808
1460	68,748	56,609	290.542	2550	134,368	113,166	324.026
1480	69,911	57,606	291.333	2600	137,449	115,832	325.222
1500	71,078	58,606	292.114	2650	140,533	118,500	326.396
1520	72,246	59,609	292.888	2700	143,620	121,172	327.549
1540	73,417	60,613	292.654	2750	146,713	123,849	328.684
1560	74,590	61,620	294.411	2800	149,808	126,528	329.800
1580	76,767	62,630	295.161	2850	152,908	129,212	330.896
1600	76,944	63,741	295.901	2900	156,009	131,898	331.975
1620	78,123	64,653	296.632	2950	159,117	134,589	333.037
1640	79,303	65,668	297.356	3000	162,226	137,283	334.084
1660	80,486	66,592	298.072	3050	165,341	139,982	335.114
1680	81,670	67,702	298.781	3100	168,456	142,681	336.126
1700	82,856	68,721	299.482	3150	171,576	145,385	337.124
1720	84,043	69,742	300.177	3200	174,695	148,089	338.109
1740	85,231	70,764	300.863	3250	177,822	150,801	339.069

TABLE A-22 Ideal Gas Properties of Hydrogen, H_2

$T(K)$, $\bar{h}$ and $\bar{u}(kJ/kmol)$, $\bar{s}°(kJ/kmol \cdot K)$

$[\bar{h}_f° = 0 \text{ kJ/kmol}]$

T	$\bar{h}$	$\bar{u}$	$\bar{s}°$	T	$\bar{h}$	$\bar{u}$	$\bar{s}°$
0	0	0	0	1440	42,808	30,835	177.410
260	7,370	5,209	126.636	1480	44,091	31,786	178.291
270	7,657	5,412	127.719	1520	45,384	32,746	179.153
280	7,945	5,617	128.765	1560	46,683	33,713	179.995
290	8,233	5,822	129.775	1600	47,990	34,687	180.820
298	8,468	5,989	130.574	1640	49,303	35,668	181.632
300	8,522	6,027	130.754	1680	50,662	36,654	182.428
320	9,100	6,440	132.621	1720	51,947	37,646	183.208
340	9,680	6,853	134.378	1760	53,279	38,645	183.973
360	10,262	7,268	136.039	1800	54,618	39,652	184.724
380	10,843	7,684	137.612	1840	55,962	40,663	185.463
400	11,426	8,100	139.106	1880	57,311	41,680	186.190
420	12,010	8,518	140.529	1920	58,668	42,705	186.904
440	12,594	8,936	141.888	1960	60,031	43,735	187.607
460	13,179	9,355	143.187	2000	61,400	44,771	188.297
480	13,764	9,773	144.432	2050	63,119	46,074	189.148
500	14,350	10,193	145.628	2100	64,847	47,386	189.979
520	14,935	10,611	146.775	2150	66,584	48,708	190.796
560	16,107	11,451	148.945	2200	68,328	50,037	191.598
600	17,280	12,291	150.968	2250	70,080	51,373	192.385
640	18,453	13,133	152.863	2300	71,839	52,716	193.159
680	19,630	13,976	154.645	2350	73,608	54,069	193.921
720	20,807	14,821	156.328	2400	75,383	55,429	194.669
760	21,988	15,669	157.923	2450	77,168	56,798	195.403
800	23,171	16,520	159.440	2500	78,960	58,175	196.125
840	24,359	17,375	160.891	2550	80,755	59,554	196.837
880	25,551	18,235	162.277	2600	82,558	60,941	197.539
920	26,747	19,098	163.607	2650	84,386	62,335	198.229
960	27,948	19,966	164.884	2700	86,186	63,737	198.907
1000	29,154	20,839	166.114	2750	88,008	65,144	199.575
1040	30,364	21,717	167.300	2800	89,838	66,558	200.234
1080	31,580	22,601	168.449	2850	91,671	67,976	200.885
1120	32,802	23,490	169.560	2900	93,512	69,401	201.527
1160	34,028	24,384	170.636	2950	95,358	70,831	202.157
1200	35,262	25,284	171.682	3000	97,211	72,268	202.778
1240	36,502	26,192	172.698	3050	99,065	73,707	203.391
1280	37,749	27,106	173.687	3100	100,926	75,152	203.995
1320	39,002	28,027	174.652	3150	102,793	76,604	204.592
1360	40,263	28,955	175.593	3200	104,667	78,061	205.181
1400	41,530	29,889	176.510	3250	106,545	79,523	205.765

TABLE A-23 Constants for the van der Waals, Redlich–Kwong, and Benedict–Webb–Rubin Equations of State

1. van der Waals and Redlich-Kwong: Constants for pressure in bars, specific volume in $m^3/kmol$, and temperature in K

Substance	van der Waals		Redlich-Kwong	
	a $bar\left(\dfrac{m^3}{kmol}\right)^2$	b $\dfrac{m^3}{kmol}$	a $bar\left(\dfrac{m^3}{kmol}\right)^2 K^{1/2}$	b $\dfrac{m^3}{kmol}$
Air	1.368	0.0367	15.989	0.02541
Butane (C_4H_{10})	13.86	0.1162	289.55	0.08060
Carbon dioxide (CO_2)	3.647	0.0428	64.43	0.02963
Carbon monoxide (CO)	1.474	0.0395	17.22	0.02737
Methane (CH_4)	2.293	0.0428	32.11	0.02965
Nitrogen (N_2)	1.366	0.0386	15.53	0.02677
Oxygen (O_2)	1.369	0.0317	17.22	0.02197
Propane (C_3H_8)	9.349	0.0901	182.23	0.06242
Refrigerant 12	10.49	0.0971	208.59	0.06731
Sulfur dioxide (SO_2)	6.883	0.0569	144.80	0.03945
Water (H_2O)	5.531	0.0305	142.59	0.02111

Source: Calculated from critical data.

2. Benedict-Webb-Rubin: Constants for pressure in bars, specific volume in $m^3/kmol$, and temperature in K

Substance	a	A	b	B	c	C	α	γ
C_4H_{10}	1.9073	10.218	0.039998	0.12436	3.206×10^5	1.006×10^6	1.101×10^{-3}	0.0340
CO_2	0.1386	2.7737	0.007210	0.04991	1.512×10^4	1.404×10^5	8.47×10^{-5}	0.00539
CO	0.0371	1.3590	0.002632	0.05454	1.054×10^3	8.676×10^3	1.350×10^{-4}	0.0060
CH_4	0.0501	1.8796	0.003380	0.04260	2.579×10^3	2.287×10^4	1.244×10^{-4}	0.0060
N_2	0.0254	1.0676	0.002328	0.04074	7.381×10^2	8.166×10^3	1.272×10^{-4}	0.0053

Source: H. W. Cooper, and J. C. Goldfrank, *Hydrocarbon Processing, 46* (12): 141 (1967).

TABLE A-24 Enthalpy of Formation, Gibbs Function of Formation, and Absolute Entropy of Various Substances at 298 K and 1 atm

$\bar{h}_f^\circ$ and $\bar{g}_f^\circ$ (kJ/kmol), $\bar{s}^\circ$ (kJ/kmol · K)

Substance	Formula	$\bar{h}_f^\circ$	$\bar{g}_f^\circ$	$\bar{s}^\circ$
Carbon	C(s)	0	0	5.74
Hydrogen	$H_2(g)$	0	0	130.57
Nitrogen	$N_2(g)$	0	0	191.50
Oxygen	$O_2(g)$	0	0	205.03
Carbon monoxide	CO(g)	−110,530	−137,150	197.54
Carbon dioxide	$CO_2(g)$	−393,520	−394,380	213.69
Water	$H_2O(g)$	−241,820	−228,590	188.72
Water	$H_2O(l)$	−285,830	−237,180	69.95
Hydrogen peroxide	$H_2O_2(g)$	−136,310	−105,600	232.63
Ammonia	$NH_3(g)$	−46,190	−16,590	192.33
Oxygen	O(g)	249,170	231,770	160.95
Hydrogen	H(g)	218,000	203,290	114.61
Nitrogen	N(g)	472,680	455,510	153.19
Hydroxyl	OH(g)	39,460	34,280	183.75
Methane	$CH_4(g)$	−74,850	−50,790	186.16
Acetylene	$C_2H_2(g)$	226,730	209,170	200.85
Ethylene	$C_2H_4(g)$	52,280	68,120	219.83
Ethane	$C_2H_6(g)$	−84,680	−32,890	229.49
Propylene	$C_3H_6(g)$	20,410	62,720	266.94
Propane	$C_3H_8(g)$	−103,850	−23,490	269.91
Butane	$C_4H_{10}(g)$	−126,150	−15,710	310.03
Pentane	$C_5H_{12}(g)$	−146,440	−8,200	348.40
Octane	$C_8H_{18}(g)$	−208,450	17,320	463.67
Octane	$C_8H_{18}(l)$	−249,910	6,610	360.79
Benzene	$C_6H_6(g)$	82,930	129,660	269.20
Methyl alcohol	$CH_3OH(g)$	−200,890	−162,140	239.70
Methyl alcohol	$CH_3OH(l)$	−238,810	−166,290	126.80
Ethyl alcohol	$C_2H_5OH(g)$	−235,310	−168,570	282.59
Ethyl alcohol	$C_2H_5OH(l)$	−277,690	174,890	160.70

Sources: Adapted from K. Wark, *Thermodynamics, 4th ed.,* McGraw-Hill, New York, 1983, as based on JANAF Thermochemical Tables, NSRDS-NBS-37, 1971; *Selected Values of Chemical Thermodynamic Properties,* NBS Tech. Note 270-3, 1968; and *API Research Project 44,* Carnegie Press, 1953.

TABLE A-25 Logarithms to the Base 10 of the Equilibrium Constant K

Temp. K	$H_2 \rightleftharpoons 2H$	$O_2 \rightleftharpoons 2O$	$N_2 \rightleftharpoons 2N$	$\frac{1}{2}O_2 + \frac{1}{2}N_2 \rightleftharpoons NO$	$H_2O \rightleftharpoons H_2 + \frac{1}{2}O_2$	$H_2O \rightleftharpoons OH + \frac{1}{2}H_2$	$CO_2 \rightleftharpoons CO + \frac{1}{2}O_2$	$CO_2 + H_2 \rightleftharpoons CO + H_2O$	Temp. °R
298	−71.224	−81.208	−159.600	−15.171	−40.048	−46.054	−45.066	−5.018	537
500	−40.316	−45.880	−92.672	−8.783	−22.886	−26.130	−25.025	−2.139	900
1000	−17.292	−19.614	−43.056	−4.062	−10.062	−11.280	−10.221	−0.159	1800
1200	−13.414	−15.208	−34.754	−3.275	−7.899	−8.811	−7.764	+0.135	2160
1400	−10.630	−12.054	−28.812	−2.712	−6.347	−7.021	−6.014	+0.333	2520
1600	−8.532	−9.684	−24.350	−2.290	−5.180	−5.677	−4.706	+0.474	2880
1700	−7.666	−8.706	−22.512	−2.116	−4.699	−5.124	−4.169	+0.530	3060
1800	−6.896	−7.836	−20.874	−1.962	−4.270	−4.613	−3.693	+0.577	3240
1900	−6.204	−7.058	−19.410	−1.823	−3.886	−4.190	−3.267	+0.619	3420
2000	−5.580	−6.356	−18.092	−1.699	−3.540	−3.776	−2.884	+0.656	3600
2100	−5.016	−5.720	−16.898	−1.586	−3.227	−3.434	−2.539	+0.688	3780
2200	−4.502	−5.142	−15.810	−1.484	−2.942	−3.091	−2.226	+0.716	3960
2300	−4.032	−4.614	−14.818	−1.391	−2.682	−2.809	−1.940	+0.742	4140
2400	−3.600	−4.130	−13.908	−1.305	−2.443	−2.520	−1.679	+0.764	4320
2500	−3.202	−3.684	−13.070	−1.227	−2.224	−2.270	−1.440	+0.784	4500
2600	−2.836	−3.272	−12.298	−1.154	−2.021	−2.038	−1.219	+0.802	4680
2700	−2.494	−2.892	−11.580	−1.087	−1.833	−1.823	−1.015	+0.818	4860
2800	−2.178	−2.536	−10.914	−1.025	−1.658	−1.624	−0.825	+0.833	5040
2900	−1.882	−2.206	−10.294	−0.967	−1.495	−1.438	−0.649	+0.846	5220
3000	−1.606	−1.898	−9.716	−0.913	−1.343	−1.265	−0.485	+0.858	5400
3100	−1.348	−1.610	−9.174	−0.863	−1.201	−1.103	−0.332	+0.869	5580
3200	−1.106	−1.340	−8.664	−0.815	−1.067	−0.951	−0.189	+0.878	5760
3300	−0.878	−1.086	−8.186	−0.771	−0.942	−0.809	−0.054	+0.888	5940
3400	−0.664	−0.846	−7.736	−0.729	−0.824	−0.674	+0.071	+0.895	6120
3500	−0.462	−0.620	−7.312	−0.690	−0.712	−0.547	+0.190	+0.902	6300

Source: Based on data from the JANAF Thermochemical Tables, NSRDS-NBS-37, 1971.

INDEX TO TABLES IN ENGLISH UNITS

TABLE A-1E Atomic or Molecular Weights and Critical Properties of Some Common Elements and Compounts

Substance	Chemical formula	M	T_c, °R	p_c, atm	$Z_c = \dfrac{p_c v_c}{RT_c}$
Acetylene	C_2H_2	26.04	556	62	0.274
Air (equivalent)	—	28.97	239	37.2	0.284
Ammonia	NH_3	17.04	730	111.3	0.242
Argon	Ar	39.94	272	47.97	0.290
Benzene	C_6H_6	78.11	1013	48.7	0.274
Butane	C_4H_{10}	58.12	765	37.5	0.274
Carbon	C	12.01	—	—	—
Carbon dioxide	CO_2	44.01	548	72.9	0.276
Carbon monoxide	CO	28.01	239	34.5	0.294
Copper	Cu	63.54	—	—	—
Ethane	C_2H_6	30.07	549	48.2	0.285
Ethyl alcohol	C_2H_5OH	46.07	929	63.0	0.249
Ethylene	C_2H_4	28.05	510	50.5	0.270
Helium	He	4.003	9.33	2.26	0.300
Hydrogen	H_2	2.018	59.8	12.8	0.304
Methane	CH_4	16.04	344	45.8	0.290
Methyl alcohol	CH_3OH	32.05	924	78.5	0.220
Nitrogen	N_2	28.01	227	33.5	0.291
Octane	C_8H_{18}	114.22	1025	24.6	0.258
Oxygen	O_2	32.00	278	49.8	0.290
Propane	C_3H_8	44.09	666	42.1	0.276
Propylene	C_3H_6	42.08	657	45.6	0.276
Refrigerant 12	CCl_2F_2	120.92	693	40.6	0.278
Refrigerant 134a	CF_3CH_2F	102.03	673	40.2	0.260
Sulfur dioxide	SO_2	64.06	775	77.7	0.268
Water	H_2O	18.02	1165	218.0	0.233

Sources: Adapted from *International Critical Tables* and L. C. Nelson and E. F. Obert, Generalized Compressibility Charts, *Chem. Eng.*, *61:* 203 (1954).

TABLE A-2E Properties of Saturated Water (Liquid–Vapor): Temperature Table

| Temp., °F | Press., lbf/in.² | Specific volume ft³/lb | | Internal energy Btu/lb | | Enthalpy Btu/lb | | | Entropy Btu/lb · °R | | Temp. °F |
		Sat. liquid v_f	Sat. vapor v_g	Sat. liquid u_f	Sat. vapor u_g	Sat. liquid h_f	Evap. h_{fg}	Sat. vapor h_g	Sat. liquid s_f	Sat. vapor s_g	
32	0.0886	0.01602	3305	−.01	1021.2	−.01	1075.4	1075.4	−.00003	2.1870	32
35	0.0999	0.01602	2948	2.99	1022.2	3.00	1073.7	1076.7	0.00607	2.1764	35
40	0.1217	0.01602	2445	8.02	1023.9	8.02	1070.9	1078.9	0.01617	2.1592	40
45	0.1475	0.01602	2037	13.04	1025.5	13.04	1068.1	1081.1	0.02618	2.1423	45
50	0.1780	0.01602	1704	18.06	1027.2	18.06	1065.2	1083.3	0.03607	2.1259	50
52	0.1917	0.01603	1589	20.06	1027.8	20.07	1064.1	1084.2	0.04000	2.1195	52
54	0.2064	0.01603	1482	22.07	1028.5	22.07	1063.0	1085.1	0.04391	2.1131	54
56	0.2219	0.01603	1383	24.08	1029.1	24.08	1061.9	1085.9	0.04781	2.1068	56
58	0.2386	0.01603	1292	26.08	1029.8	26.08	1060.7	1086.8	0.05159	2.1005	58
60	0.2563	0.01604	1207	28.08	1030.4	28.08	1059.6	1087.7	0.05555	2.0943	60
62	0.2751	0.01604	1129	30.09	1031.1	30.09	1058.5	1088.6	0.05940	2.0882	62
64	0.2952	0.01604	1056	32.09	1031.8	32.09	1057.3	1089.4	0.06323	2.0821	64
66	0.3165	0.01604	988.4	34.09	1032.4	34.09	1056.2	1090.3	0.06704	2.0761	66
68	0.3391	0.01605	925.8	36.09	1033.1	36.09	1055.1	1091.2	0.07084	2.0701	68
70	0.3632	0.01605	867.7	38.09	1033.7	38.09	1054.0	1092.0	0.07463	2.0642	70
72	0.3887	0.01606	813.7	40.09	1034.4	40.09	1052.8	1092.9	0.07839	2.0584	72
74	0.4158	0.01606	763.5	42.09	1035.0	42.09	1051.7	1093.8	0.08215	2.0526	74
76	0.4446	0.01606	716.8	44.09	1035.7	44.09	1050.6	1094.7	0.08589	2.0469	76
78	0.4750	0.01607	673.3	46.09	1036.3	46.09	1049.4	1095.5	0.08961	2.0412	78
80	0.5073	0.01607	632.8	48.08	1037.0	48.09	1048.3	1096.4	0.09332	2.0356	80
82	0.5414	0.01608	595.0	50.08	1037.6	50.08	1047.2	1097.3	0.09701	2.0300	82
84	0.5776	0.01608	559.8	52.08	1038.3	52.08	1046.0	1098.1	0.1007	2.0245	84
86	0.6158	0.01609	527.0	54.08	1038.9	54.08	1044.9	1099.0	0.1044	2.0190	86
88	0.6562	0.01609	496.3	56.07	1039.6	56.07	1043.8	1099.9	0.1080	2.0136	88
90	0.6988	0.01610	467.7	58.07	1040.2	58.07	1042.7	1100.7	0.1117	2.0083	90
92	0.7439	0.01611	440.9	60.06	1040.9	60.06	1041.5	1101.6	0.1153	2.0030	92
94	0.7914	0.01611	415.9	62.06	1041.5	62.06	1040.4	1102.4	0.1189	1.9977	94
96	0.8416	0.01612	392.4	64.05	1041.2	64.06	1039.2	1103.3	0.1225	1.9925	96
98	0.8945	0.01612	370.5	66.05	1042.8	66.05	1038.1	1104.2	0.1261	1.9874	98
100	0.9503	0.01613	350.0	68.04	1043.5	68.05	1037.0	1105.0	0.1296	1.9822	100
110	1.276	0.01617	265.1	78.02	1046.7	78.02	1031.3	1109.3	0.1473	1.9574	110
120	1.695	0.01621	203.0	87.99	1049.9	88.00	1025.5	1113.5	0.1647	1.9336	120
130	2.225	0.01625	157.2	97.97	1053.0	97.98	1019.8	1117.8	0.1817	1.9109	130
140	2.892	0.01629	122.9	107.95	1056.2	107.96	1014.0	1121.9	0.1985	1.8892	140
150	3.722	0.01634	97.0	117.95	1059.3	117.96	1008.1	1126.1	0.2150	1.8684	150
160	4.745	0.01640	77.2	127.94	1062.3	127.96	1002.2	1130.1	0.2313	1.8484	160
170	5.996	0.01645	62.0	137.95	1065.4	137.97	996.2	1134.2	0.2473	1.8293	170
180	7.515	0.01651	50.2	147.97	1068.3	147.99	990.2	1138.2	0.2631	1.8109	180
190	9.343	0.01657	41.0	158.00	1071.3	158.03	984.1	1142.1	0.2787	1.7932	190
200	11.529	0.01663	33.6	168.04	1074.2	168.07	977.9	1145.9	0.2940	1.7762	200

TABLE A-2E (*Continued*)

Temp. °F	Press. lbf/in.²	Specific volume ft³/lb		Internal energy Btu/lb		Enthalpy Btu/lb			Entropy Btu/lb · °R		Temp. °F
		Sat. liquid v_f	Sat. vapor v_g	Sat. liquid u_f	Sat. vapor u_g	Sat. liquid h_f	Evap. h_{fg}	Sat. vapor h_g	Sat. liquid s_f	Sat. vapor s_g	
210	14.13	0.01670	27.82	178.1	1077.0	178.1	971.6	1149.7	0.3091	1.7599	210
212	14.70	0.01672	26.80	180.1	1077.6	180.2	970.3	1150.5	0.3121	1.7567	212
220	17.19	0.01677	23.15	188.2	1079.8	188.2	965.3	1153.5	0.3241	1.7441	220
230	20.78	0.01685	19.39	198.3	1082.6	198.3	958.8	1157.1	0.3388	1.7289	230
240	24.97	0.01692	16.33	208.4	1085.3	208.4	952.3	1160.7	0.3534	1.7143	240
250	29.82	0.01700	13.83	218.5	1087.9	218.6	945.6	1164.2	0.3677	1.7001	250
260	35.42	0.01708	11.77	228.6	1090.5	228.8	938.8	1167.6	0.3819	1.6864	260
270	41.85	0.01717	10.07	238.8	1093.0	239.0	932.0	1170.9	0.3960	1.6731	270
280	49.18	0.01726	8.65	249.0	1095.4	249.2	924.9	1174.1	0.4099	1.6602	280
290	57.53	0.01735	7.47	259.3	1097.7	259.4	917.8	1177.2	0.4236	1.6477	290
300	66.98	0.01745	6.472	269.5	1100.0	269.7	910.4	1180.2	0.4372	1.6356	300
310	77.64	0.01755	5.632	279.8	1102.1	280.1	903.0	1183.0	0.4507	1.6238	310
320	89.60	0.01765	4.919	290.1	1104.2	290.4	895.3	1185.8	0.4640	1.6123	320
330	103.00	0.01776	4.312	300.5	1106.2	300.8	887.5	1188.4	0.4772	1.6010	330
340	117.93	0.01787	3.792	310.9	1108.0	311.3	879.5	1190.8	0.4903	1.5901	340
350	134.53	0.01799	3.346	321.4	1109.8	321.8	871.3	1193.1	0.5033	1.5793	350
360	152.92	0.01811	2.961	331.8	1111.4	332.4	862.9	1195.2	0.5162	1.5688	360
370	173.23	0.01823	2.628	342.4	1112.9	343.0	854.2	1197.2	0.5289	1.5585	370
380	195.60	0.01836	2.339	353.0	1114.3	353.6	845.4	1199.0	0.5416	1.5483	380
390	220.2	0.01850	2.087	363.6	1115.6	364.3	836.2	1200.6	0.5542	1.5383	390
400	247.1	0.01864	1.866	374.3	1116.6	375.1	826.8	1202.0	0.5667	1.5284	400
410	276.5	0.01878	1.673	385.0	1117.6	386.0	817.2	1203.1	0.5792	1.5187	410
420	308.5	0.01894	1.502	395.8	1118.3	396.9	807.2	1204.1	0.5915	1.5091	420
430	343.3	0.01909	1.352	406.7	1118.9	407.9	796.9	1204.8	0.6038	1.4995	430
440	381.2	0.01926	1.219	417.6	1119.3	419.0	786.3	1205.3	0.6161	1.4900	440
450	422.1	0.01943	1.1011	428.6	1119.5	430.2	775.4	1205.6	0.6282	1.4806	450
460	466.3	0.01961	0.9961	439.7	1119.6	441.4	764.1	1205.5	0.6404	1.4712	460
470	514.1	0.01980	0.9025	450.9	1119.4	452.8	752.4	1205.2	0.6525	1.4618	470
480	565.5	0.02000	0.8187	462.2	1118.9	464.3	740.3	1204.6	0.6646	1.4524	480
490	620.7	0.02021	0.7436	473.6	1118.3	475.9	727.8	1203.7	0.6767	1.4430	490
500	680.0	0.02043	0.6761	485.1	1117.4	487.7	714.8	1202.5	0.6888	1.4335	500
520	811.4	0.02091	0.5605	508.5	1114.8	511.7	687.3	1198.9	0.7130	1.4145	520
540	961.5	0.02145	0.4658	532.6	1111.0	536.4	657.5	1193.8	0.7374	1.3950	540
560	1131.8	0.02207	0.3877	548.4	1105.8	562.0	625.0	1187.0	0.7620	1.3749	560
580	1324.3	0.02278	0.3225	583.1	1098.9	588.6	589.3	1178.0	0.7872	1.3540	580
600	1541.0	0.02363	0.2677	609.9	1090.0	616.7	549.7	1166.4	0.8130	1.3317	600
620	1784.4	0.02465	0.2209	638.3	1078.5	646.4	505.0	1151.4	0.8398	1.3075	620
640	2057.1	0.02593	0.1805	668.7	1063.2	678.6	453.4	1131.9	0.8681	1.2803	640
660	2362	0.02767	0.1446	702.3	1042.3	714.4	391.1	1105.5	0.8990	1.2483	660
680	2705	0.03032	0.1113	741.7	1011.0	756.9	309.8	1066.7	0.9350	1.2068	680
700	3090	0.03666	0.0744	801.7	947.7	822.7	167.5	990.2	0.9902	1.1346	700
705.4	3204	0.05053	0.05053	872.6	872.6	902.5	0	902.5	1.0580	1.0580	705.4

Source: Tables A-2E through A-5E are adapted from K. Wark, *Thermodynamics, 4th ed.,* McGraw-Hill, New York, 1983, as extracted from J. H. Keenan, F. G. Keyes, P. G. Hill, and J. G. Moore, "Steam Tables," Wiley, New York, 1969.

TABLE A-3E Properties of Saturated Water (Liquid–Vapor): Pressure Table

Press. lbf/in.^{2}	Temp., °F	Specific volume ft^3/lb		Internal energy Btu/lb		Enthalpy Btu/lb			Entropy Btu/lb · °R			Press. lbf/in.^{2}
		Sat. liquid v_f	Sat. vapor v_g	Sat. liquid u_f	Sat. vapor u_g	Sat. liquid h_f	Evap. h_{fg}	Sat. vapor h_g	Sat. liquid s_f	Evap. s_{fg}	Sat. vapor s_g	
0.4	72.84	0.01606	792.0	40.94	1034.7	40.94	1052.3	1093.3	0.0800	1.9760	2.0559	0.4
0.6	85.19	0.01609	540.0	53.26	1038.7	53.27	1045.4	1098.6	0.1029	1.9184	2.0213	0.6
0.8	94.35	0.01611	411.7	62.41	1041.7	62.41	1040.2	1102.6	0.1195	1.8773	1.9968	0.8
1.0	101.70	0.01614	333.6	69.74	1044.0	69.74	1036.0	1105.8	0.1327	1.8453	1.9779	1.0
1.2	107.88	0.01616	280.9	75.90	1046.0	75.90	1032.5	1108.4	0.1436	1.8190	1.9626	1.2
1.5	115.65	0.01619	227.7	83.65	1048.5	83.65	1028.0	1111.7	0.1571	1.7867	1.9438	1.5
2.0	126.04	0.01623	173.75	94.02	1051.8	94.02	1022.1	1116.1	0.1750	1.7448	1.9198	2.0
3.0	141.43	0.01630	118.72	109.38	1056.6	109.39	1013.1	1122.5	0.2009	1.6852	1.8861	3.0
4.0	152.93	0.01636	90.64	120.88	1060.2	120.89	1006.4	1127.3	0.2198	1.6426	1.8624	4.0
5.0	162.21	0.01641	73.53	130.15	1063.0	130.17	1000.9	1131.0	0.2349	1.6093	1.8441	5.0
6.0	170.03	0.01645	61.98	137.98	1065.4	138.00	996.2	1134.2	0.2474	1.5819	1.8292	6.0
7.0	176.82	0.01649	53.65	144.78	1067.4	144.80	992.1	1136.9	0.2581	1.5585	1.8167	7.0
8.0	182.84	0.01653	47.35	150.81	1069.2	150.84	988.4	1139.3	0.2675	1.5383	1.8058	8.0
9.0	188.26	0.01656	42.41	156.25	1070.8	156.27	985.1	1141.4	0.2760	1.5203	1.7963	9.0
10	193.19	0.01659	38.42	161.20	1072.2	161.23	982.1	1143.3	0.2836	1.5041	1.7877	10
14.696	211.99	0.01672	26.80	180.10	1077.6	180.15	970.4	1150.5	0.3121	1.4446	1.7567	14.696
15	213.03	0.01672	26.29	181.14	1077.9	181.19	969.7	1150.9	0.3137	1.4414	1.7551	15
20	227.96	0.01683	20.09	196.19	1082.0	196.26	960.1	1156.4	0.3358	1.3962	1.7320	20
25	240.08	0.01692	16.31	208.44	1085.3	208.52	952.2	1160.7	0.3535	1.3607	1.7142	25
30	250.34	0.01700	13.75	218.84	1088.0	218.93	945.4	1164.3	0.3682	1.3314	1.6996	30
35	259.30	0.01708	11.90	227.93	1090.3	228.04	939.3	1167.4	0.3809	1.3064	1.6873	35
40	267.26	0.01715	10.50	236.03	1092.3	236.16	933.8	1170.0	0.3921	1.2845	1.6767	40
45	274.46	0.01721	9.40	243.37	1094.0	243.51	928.8	1172.3	0.4022	1.2651	1.6673	45
50	281.03	0.01727	8.52	250.08	1095.6	250.24	924.2	1174.4	0.4113	1.2476	1.6589	50
55	287.10	0.01733	7.79	256.28	1097.0	256.46	919.9	1176.3	0.4196	1.2317	1.6513	55
60	292.73	0.01738	7.177	262.1	1098.3	262.2	915.8	1178.0	0.4273	1.2170	1.6443	60
65	298.00	0.01743	6.647	267.5	1099.5	267.7	911.9	1179.6	0.4345	1.2035	1.6380	65
70	302.96	0.01748	6.209	272.6	1100.6	272.8	908.3	1181.0	0.4412	1.1909	1.6321	70
75	307.63	0.01752	5.818	277.4	1101.6	277.6	904.8	1182.4	0.4475	1.1790	1.6265	75
80	312.07	0.01757	5.474	282.0	1102.6	282.2	901.4	1183.6	0.4534	1.1679	1.6213	80
85	316.29	0.01761	5.170	286.3	1103.5	286.6	898.2	1184.8	0.4591	1.1574	1.6165	85
90	320.31	0.01766	4.898	290.5	1104.3	290.8	895.1	1185.9	0.4644	1.1475	1.6119	90
95	324.16	0.01770	4.654	294.5	1105.0	294.8	892.1	1186.9	0.4695	1.1380	1.6075	95
100	327.86	0.01774	4.434	298.3	1105.8	298.6	889.2	1187.8	0.4744	1.1290	1.6034	100
110	334.82	0.01781	4.051	305.5	1107.1	305.9	883.7	1189.6	0.4836	1.1122	1.5958	110
120	341.30	0.01789	3.730	312.3	1108.3	312.7	878.5	1191.1	0.4920	1.0966	1.5886	120
130	347.37	0.01797	3.457	318.6	1109.4	319.0	873.5	1192.5	0.4999	1.0822	1.5821	130
140	353.08	0.01802	3.221	324.6	1110.3	325.1	868.7	1193.8	0.5073	1.0688	1.5761	140
150	358.48	0.01809	3.016	330.2	1111.2	330.8	864.2	1194.9	0.5142	1.0562	1.5704	150
160	363.60	0.01815	2.836	335.6	1112.0	336.2	859.8	1196.0	0.5208	1.0443	1.5651	160

TABLE A-3E (*Continued*)

Press. lbf/in.²	Temp., °F	Specific volume ft³/lb		Internal energy Btu/lb		Enthalpy Btu/lb			Entropy Btu/lb · °R			Press. lbf/in.²
		Sat. liquid v_f	Sat. vapor v_g	Sat. liquid u_f	Sat. vapor u_g	Sat. liquid h_f	Evap. h_{fg}	Sat. vapor h_g	Sat. liquid s_f	Evap. s_{fg}	Sat. vapor s_g	
170	368.47	0.01821	2.676	340.8	1112.7	341.3	855.6	1196.9	0.5270	1.0330	1.5600	170
180	373.13	0.01827	2.553	345.7	1113.4	346.3	851.5	1197.8	0.5329	1.0223	1.5552	180
190	377.59	0.01833	2.405	350.4	1114.0	351.0	847.5	1198.6	0.5386	1.0122	1.5508	190
200	381.86	0.01839	2.289	354.9	1114.6	355.6	843.7	1199.3	0.5440	1.0025	1.5465	200
250	401.04	0.01865	1.845	375.4	1116.7	376.2	825.8	1202.1	0.5680	0.9594	1.5274	250
300	417.43	0.01890	1.544	393.0	1118.2	394.1	809.8	1203.9	0.5883	0.9232	1.5115	300
350	431.82	0.01912	1.327	408.7	1119.0	409.9	795.0	1204.9	0.6060	0.8917	1.4977	350
400	444.70	0.01934	1.162	422.8	1119.5	424.2	781.2	1205.5	0.6218	0.8638	1.4856	400
450	456.39	0.01955	1.033	435.7	1119.6	437.4	768.2	1205.6	0.6360	0.8385	1.4745	450
500	467.13	0.01975	0.928	447.7	1119.4	449.5	755.8	1205.3	0.6490	0.8154	1.4644	500
550	477.07	0.01994	0.842	458.9	1119.1	460.9	743.9	1204.8	0.6611	0.7941	1.4551	550
600	486.33	0.02013	0.770	469.4	1118.6	471.7	732.4	1204.1	0.6723	0.7742	1.4464	600
700	503.23	0.02051	0.656	488.9	1117.0	491.5	710.5	1202.0	0.6927	0.7378	1.4305	700
800	518.36	0.02087	0.569	506.6	1115.0	509.7	689.6	1199.3	0.7110	0.7050	1.4160	800
900	532.12	0.02123	0.501	523.0	1112.6	526.6	669.5	1196.0	0.7277	0.6750	1.4027	900
1000	544.75	0.02159	0.446	538.4	1109.9	542.4	650.0	1192.4	0.7432	0.6471	1.3903	1000
1100	556.45	0.02195	0.401	552.9	1106.8	557.4	631.0	1188.3	0.7576	0.6209	1.3786	1100
1200	567.37	0.02232	0.362	566.7	1103.5	571.7	612.3	1183.9	0.7712	0.5961	1.3673	1200
1300	577.60	0.02269	0.330	579.9	1099.8	585.4	593.8	1179.2	0.7841	0.5724	1.3565	1300
1400	587.25	0.02307	0.302	592.7	1096.0	598.6	575.5	1174.1	0.7964	0.5497	1.3461	1400
1500	596.39	0.02346	0.277	605.0	1091.8	611.5	557.2	1168.7	0.8082	0.5276	1.3359	1500
1600	605.06	0.02386	0.255	616.9	1087.4	624.0	538.9	1162.9	0.8196	0.5062	1.3258	1600
1700	613.32	0.02428	0.236	628.6	1082.7	636.2	520.6	1156.9	0.8307	0.4852	1.3159	1700
1800	621.21	0.02472	0.218	640.0	1077.7	648.3	502.1	1150.4	0.8414	0.4645	1.3060	1800
1900	628.76	0.02517	0.203	651.3	1072.3	660.1	483.4	1143.5	0.8519	0.4441	1.2961	1900
2000	636.00	0.02565	0.188	662.4	1066.6	671.9	464.4	1136.3	0.8623	0.4238	1.2861	2000
2250	652.90	0.02698	0.157	689.9	1050.6	701.1	414.8	1115.9	0.8876	0.3728	1.2604	2250
2500	668.31	0.02860	0.131	717.7	1031.0	730.9	360.5	1091.4	0.9131	0.3196	1.2327	2500
2750	682.46	0.03077	0.107	747.3	1005.9	763.0	297.4	1060.4	0.9401	0.2604	1.2005	2750
3000	695.52	0.03431	0.084	783.4	968.8	802.5	213.0	1015.5	0.9732	0.1843	1.1575	3000
3203.6	705.44	0.05053	0.0505	872.6	872.6	902.5	0	902.5	1.0580	0	1.0580	3203.6

TABLE A-4E Properties of Superheated Water Vapor

T °F	v ft³/lb	u Btu/lb	h Btu/lb	s Btu/lb·°R	v ft³/lb	u Btu/lb	h Btu/lb	s Btu/lb·°R
	$p = 1$ lbf/in.² ($T_{sat} = 101.7°F$)				$p = 5$ lbf/in.² ($T_{sat} = 162.2°F$)			
Sat.	333.6	1044.0	1105.8	1.9779	73.53	1063.0	1131.0	1.8441
150	362.6	1060.4	1127.5	2.0151				
200	392.5	1077.5	1150.1	2.0508	78.15	1076.0	1148.6	1.8715
250	422.4	1094.7	1172.8	2.0839	84.21	1093.8	1171.7	1.9052
300	452.3	1112.0	1195.7	2.1150	90.24	1111.3	1194.8	1.9367
400	511.9	1147.0	1241.8	2.1720	102.24	1146.6	1241.2	1.9941
500	571.5	1182.8	1288.5	2.2235	114.20	1182.5	1288.2	2.0458
600	631.1	1219.3	1336.1	2.2706	126.15	1219.1	1335.8	2.0930
700	690.7	1256.7	1384.5	2.3142	138.08	1256.5	1384.3	2.1367
800	750.3	1294.4	1433.7	2.3550	150.01	1294.7	1433.5	2.1775
900	809.9	1333.9	1483.8	2.3932	161.94	1333.8	1483.7	2.2158
1000	869.5	1373.9	1534.8	2.4294	173.86	1373.9	1534.7	2.2520
	$p = 10$ lbf/in.² ($T_{sat} = 193.2°F$)				$p = 14.7$ lbf/in.² ($T_{sat} = 212.0°F$)			
Sat.	38.42	1072.2	1143.3	1.7877	26.80	1077.6	1150.5 _1143.2_	1.7567
200	38.85	1074.7	1146.6	1.7927				
250	41.95	1092.6	1170.2	1.8272	28.42	1091.5	1168.8	1.7832
300	44.99	1110.4	1193.7	1.8592	30.52	1109.6	1192.6	1.8157
400	51.03	1146.1	1240.5	1.9171	34.67	1145.6	1239.9	1.8741
500	57.04	1182.2	1287.7	1.9690	38.77	1181.8	1287.3	1.9263
600	63.03	1218.9	1335.5	2.0164	42.86	1218.6	1335.2	1.9737
700	69.01	1256.3	1384.0	2.0601	46.93	1256.1	1383.8	2.0175
800	74.98	1294.6	1433.3	2.1009	51.00	1294.4	1433.1	2.0584
900	80.95	1333.7	1483.5	2.1393	55.07	1333.6	1483.4	2.0967
1000	86.91	1373.8	1534.6	2.1755	59.13	1373.7	1534.5	2.1330
1100	92.88	1414.7	1586.6	2.2099	63.19	1414.6	1586.4	2.1674
	$p = 20$ lbf/in.² ($T_{sat} = 228.0°F$) _1166.9_				$p = 40$ lbf/in.² ($T_{sat} = 267.3°F$)			
Sat.	20.09	1082.0	1156.4	1.7320	10.50	1093.3	1170.0	1.6767
250	20.79	1090.3	1167.2	1.7475				
300	22.36	1108.7	1191.5	1.7805	11.04	1105.1	1186.8	1.6993
350	23.90	1126.9	1215.4	1.8110	11.84	1124.2	1211.8	1.7312
400	25.43	1145.1	1239.2	1.8395	12.62	1143.0	1236.4	1.7606
500	28.46	1181.5	1286.8	1.8919	14.16	1180.1	1284.9	1.8140
600	31.47	1218.4	1334.8	1.9395	15.69	1217.3	1333.4	1.8621
700	34.47	1255.9	1383.5	1.9834	17.20	1255.1	1382.4	1.9063
800	37.46	1294.3	1432.9	2.0243	18.70	1293.7	1432.1	1.9474
900	40.45	1333.5	1483.2	2.0627	20.20	1333.0	1482.5	1.9859
1000	43.44	1373.5	1534.3	2.0989	21.70	1373.1	1533.8	2.0223
1100	46.42	1414.5	1586.3	2.1334	23.20	1414.2	1585.9	2.0568

TABLE A-4E (*Continued*)

T °F	v ft³/lb	u Btu/lb	h Btu/lb	s Btu/lb · °R	v ft³/lb	u Btu/lb	h Btu/lb	s Btu/lb · °R
	$p = 60$ lbf/in.² ($T_{sat} = 292.7°F$)				$p = 80$ lbf/in.² ($T_{sat} = 312.1°F$)			
Sat.	7.17	1098.3	1178.0	1.6444	5.47	1102.6	1183.6	1.6214
300	7.26	1101.3	1181.9	1.6496				
350	7.82	1121.4	1208.2	1.6830	5.80	1118.5	1204.3	1.6476
400	8.35	1140.8	1233.5	1.7134	6.22	1138.5	1230.6	1.6790
500	9.40	1178.6	1283.0	1.7678	7.02	1177.2	1281.1	1.7346
600	10.43	1216.3	1332.1	1.8165	7.79	1215.3	1330.7	1.7838
700	11.44	1254.4	1381.4	1.8609	8.56	1253.6	1380.3	1.8285
800	12.45	1293.0	1431.2	1.9022	9.32	1292.4	1430.4	1.8700
900	13.45	1332.5	1481.8	1.9408	10.08	1332.0	1481.2	1.9087
1000	14.45	1372.7	1533.2	1.9773	10.83	1372.3	1532.6	1.9453
1100	15.45	1413.8	1585.4	2.0119	11.58	1413.5	1584.9	1.9799
1200	16.45	1455.8	1638.5	2.0448	12.33	1455.5	1638.1	2.0130
	$p = 100$ lbf/in.² ($T_{sat} = 327.8°F$)				$p = 120$ lbf/in.² ($T_{sat} = 341.3°F$)			
Sat.	4.434	1105.8	1187.8	1.6034	3.730	1108.3	1191.1	1.5886
350	4.592	1115.4	1200.4	1.6191	3.783	1112.2	1196.2	1.5950
400	4.934	1136.2	1227.5	1.6517	4.079	1133.8	1224.4	1.6288
450	5.265	1156.2	1253.6	1.6812	4.360	1154.3	1251.2	1.6590
500	5.587	1175.7	1279.1	1.7085	4.633	1174.2	1277.1	1.6868
600	6.216	1214.2	1329.3	1.7582	5.164	1213.2	1327.8	1.7371
700	6.834	1252.8	1379.2	1.8033	5.682	1252.0	1378.2	1.7825
800	7.445	1291.8	1429.6	1.8449	6.195	1291.2	1428.7	1.8243
900	8.053	1331.5	1480.5	1.8838	6.703	1330.9	1479.8	1.8633
1000	8.657	1371.9	1532.1	1.9204	7.208	1371.5	1531.5	1.9000
1100	9.260	1413.1	1584.5	1.9551	7.711	1412.8	1584.0	1.9348
1200	9.861	1455.2	1637.7	1.9882	8.213	1454.9	1637.3	1.9679
	$p = 140$ lbf/in.² ($T_{sat} = 353.1°F$)				$p = 160$ lbf/in.² ($T_{sat} = 363.6°F$)			
Sat.	3.221	1110.3	1193.8	1.5761	2.836	1112.0	1196.0	1.5651
400	3.466	1131.4	1221.2	1.6088	3.007	1128.8	1217.8	1.5911
450	3.713	1152.4	1248.6	1.6399	3.228	1150.5	1246.1	1.6230
500	3.952	1172.7	1275.1	1.6682	3.440	1171.2	1273.0	1.6518
550	4.184	1192.5	1300.9	1.6945	3.646	1191.3	1299.2	1.6785
600	4.412	1212.1	1326.4	1.7191	3.848	1211.1	1325.0	1.7034
700	4.860	1251.2	1377.1	1.7648	4.243	1250.4	1376.0	1.7494
800	5.301	1290.5	1427.9	1.8068	4.631	1289.9	1427.0	1.7916
900	5.739	1330.4	1479.1	1.8459	5.015	1329.9	1478.4	1.8308
1000	6.173	1371.0	1531.0	1.8827	5.397	1370.6	1530.4	1.8677
1100	6.605	1412.4	1583.6	1.9176	5.776	1412.1	1583.1	1.9026
1200	7.036	1454.6	1636.9	1.9507	6.154	1454.3	1636.5	1.9358

TABLE A-4E (*Continued*)

T °F	v ft³/lb	u Btu/lb	h Btu/lb	s Btu/lb · °R	v ft³/lb	u Btu/lb	h Btu/lb	s Btu/lb · °R
	$p = 180$ lbf/in.² ($T_{sat} = 373.1°F$)				$p = 200$ lbf/in.² ($T_{sat} = 381.8°F$)			
Sat.	2.533	1113.4	1197.8	1.5553	2.289	1114.6	1199.3	1.5464
400	2.648	1126.2	1214.4	1.5749	2.361	1123.5	1210.8	1.5600
450	2.850	1148.5	1243.4	1.6078	2.548	1146.4	1240.7	1.5938
500	3.042	1169.6	1270.9	1.6372	2.724	1168.0	1268.8	1.6239
550	3.228	1190.0	1297.5	1.6642	2.893	1188.7	1295.7	1.6512
600	3.409	1210.0	1323.5	1.6893	3.058	1208.9	1322.1	1.6767
700	3.763	1249.6	1374.9	1.7357	3.379	1248.8	1373.8	1.7234
800	4.110	1289.3	1426.2	1.7781	3.693	1288.6	1425.3	1.7660
900	4.453	1329.4	1477.7	1.8174	4.003	1328.9	1477.1	1.8055
1000	4.793	1370.2	1529.8	1.8545	4.310	1369.8	1529.3	1.8425
1100	5.131	1411.7	1582.6	1.8894	4.615	1411.4	1582.2	1.8776
1200	5.467	1454.0	1636.1	1.9227	4.918	1453.7	1635.7	1.9109
	$p = 250$ lbf/in.² ($T_{sat} = 401.0°F$)				$p = 300$ lbf/in.² ($T_{sat} = 417.4°F$)			
Sat.	1.845	1116.7	1202.1	1.5274	1.544	1118.2	1203.9	1.5115
450	2.002	1141.1	1233.7	1.5632	1.636	1135.4	1226.2	1.5365
500	2.150	1163.8	1263.3	1.5948	1.766	1159.5	1257.5	1.5701
550	2.290	1185.3	1291.3	1.6233	1.888	1181.9	1286.7	1.5997
600	2.426	1206.1	1318.3	1.6494	2.004	1203.2	1314.5	1.6266
700	2.688	1246.7	1371.1	1.6970	2.227	1244.0	1368.3	1.6751
800	2.943	1287.0	1423.2	1.7301	2.442	1285.4	1421.0	1.7187
900	3.193	1327.6	1475.3	1.7799	2.653	1326.3	1473.6	1.7589
1000	3.440	1368.7	1527.9	1.8172	2.860	1367.7	1526.5	1.7964
1100	3.685	1410.5	1581.0	1.8524	3.066	1409.6	1579.8	1.8317
1200	3.929	1453.0	1634.8	1.8858	3.270	1452.2	1633.8	1.8653
1300	4.172	1496.3	1689.3	1.9177	3.473	1495.6	1688.4	1.8973
	$p = 350$ lbf/in.² ($T_{sat} = 431.8°F$)				$p = 400$ lbf/in.² ($T_{sat} = 444.7°F$)			
Sat.	1.327	1119.0	1204.9	1.4978	1.162	1119.5	1205.5	1.4856
450	1.373	1129.2	1218.2	1.5125	1.175	1122.6	1209.6	1.4901
500	1.491	1154.9	1251.5	1.5482	1.284	1150.1	1245.2	1.5282
550	1.600	1178.3	1281.9	1.5790	1.383	1174.6	1277.0	1.5605
600	1.703	1200.3	1310.6	1.6068	1.476	1197.3	1306.6	1.5892
700	1.898	1242.5	1365.4	1.6562	1.650	1240.4	1362.5	1.6397
800	2.085	1283.8	1418.8	1.7004	1.816	1282.1	1416.6	1.6844
900	2.267	1325.0	1471.8	1.7409	1.978	1323.7	1470.1	1.7252
1000	2.446	1366.6	1525.0	1.7787	2.136	1365.5	1523.6	1.7632
1100	2.624	1408.7	1578.6	1.8142	2.292	1407.8	1577.4	1.7989
1200	2.799	1451.5	1632.8	1.8478	2.446	1450.7	1631.8	1.8327
1300	2.974	1495.0	1687.6	1.8799	2.599	1494.3	1686.8	1.8648

TABLE A-4E (*Continued*)

T °F	v ft^3/lb	u Btu/lb	h Btu/lb	s Btu/lb · °R	v ft^3/lb	u Btu/lb	h Btu/lb	s Btu/lb · °R
	$p = 450$ lbf/in.^{2} ($T_{sat} = 456.4$°F)				$p = 500$ lbf/in.^{2} ($T_{sat} = 467.1$°F)			
Sat.	1.033	1119.6	1205.6	1.4746	0.928	1119.4	1205.3	1.4645
500	1.123	1145.1	1238.5	1.5097	0.992	1139.7	1231.5	1.4923
550	1.215	1170.7	1271.9	1.5436	1.079	1166.7	1266.6	1.5279
600	1.300	1194.3	1302.5	1.5732	1.158	1191.1	1298.3	1.5585
700	1.458	1238.2	1359.6	1.6248	1.304	1236.0	1356.7	1.6112
800	1.608	1280.5	1414.4	1.6701	1.441	1278.8	1412.1	1.6571
900	1.752	1322.4	1468.3	1.7113	1.572	1321.0	1466.5	1.6987
1000	1.894	1364.4	1522.2	1.7495	1.701	1363.3	1520.7	1.7471
1100	2.034	1406.9	1576.3	1.7853	1.827	1406.0	1575.1	1.7731
1200	2.172	1450.0	1630.8	1.8192	1.952	1449.2	1629.8	1.8072
1300	2.308	1493.7	1685.9	1.8515	2.075	1493.1	1685.1	1.8395
1400	2.444	1538.1	1741.7	1.8823	2.198	1537.6	1741.0	1.8704
	$p = 600$ lbf/in.^{2} ($T_{sat} = 486.3$°F)				$p = 700$ lbf/in.^{2} ($T_{sat} = 503.2$°F)			
Sat.	0.770	1118.6	1204.1	1.4464	0.656	1117.0	1202.0	1.4305
500	0.795	1128.0	1216.2	1.4592				
550	0.875	1158.2	1255.4	1.4990	0.728	1149.0	1243.2	1.4723
600	0.946	1184.5	1289.5	1.5320	0.793	1177.5	1280.2	1.5081
700	1.073	1231.5	1350.6	1.5872	0.907	1226.9	1344.4	1.5661
800	1.190	1275.4	1407.6	1.6343	1.011	1272.0	1402.9	1.6145
900	1.302	1318.4	1462.9	1.6766	1.109	1315.6	1459.3	1.6576
1000	1.411	1361.2	1517.8	1.7155	1.204	1358.9	1514.9	1.6970
1100	1.517	1404.2	1572.7	1.7519	1.296	1402.4	1570.2	1.7337
1200	1.622	1447.7	1627.8	1.7861	1.387	1446.2	1625.8	1.7682
1300	1.726	1491.7	1683.4	1.8186	1.476	1490.4	1681.7	1.8009
1400	1.829	1536.5	1739.5	1.8497	1.565	1535.3	1738.1	1.8321
	$p = 800$ lbf/in.^{2} ($T_{sat} = 518.3$°F)				$p = 900$ lbf/in.^{2} ($T_{sat} = 532.1$°F)			
Sat.	0.569	1115.0	1199.3	1.4160	0.501	1112.6	1196.0	1.4027
550	0.615	1138.8	1229.9	1.4469	0.527	1127.5	1215.2	1.4219
600	0.677	1170.1	1270.4	1.4861	0.587	1162.2	1260.0	1.4652
650	0.732	1197.2	1305.6	1.5186	0.639	1191.1	1297.5	1.4999
700	0.783	1222.1	1338.0	1.5471	0.686	1217.1	1331.4	1.5297
800	0.876	1268.5	1398.2	1.5969	0.772	1264.9	1393.4	1.5810
900	0.964	1312.9	1455.6	1.6408	0.851	1310.1	1451.9	1.6257
1000	1.048	1356.7	1511.9	1.6807	0.927	1354.5	1508.9	1.6662
1100	1.130	1400.5	1567.8	1.7178	1.001	1398.7	1565.4	1.7036
1200	1.210	1444.6	1623.8	1.7526	1.073	1443.0	1621.7	1.7386
1300	1.289	1489.1	1680.0	1.7854	1.144	1487.8	1687.3	1.7717
1400	1.367	1534.2	1736.6	1.8167	1.214	1533.0	1735.1	1.8031

TABLE A-4E (*Continued*)

T °F	v ft³/lb	u Btu/lb	h Btu/lb	s Btu/lb · °R	v ft³/lb	u Btu/lb	h Btu/lb	s Btu/lb · °R
	$p = 1000$ lbf/in.² ($T_{sat} = 544.7°F$)				$p = 1200$ lbf/in.² ($T_{sat} = 567.4°F$)			
Sat.	0.446	1109.0	1192.4	1.3903	0.362	1103.5	1183.9	1.3673
600	0.514	1153.7	1248.8	1.4450	0.402	1134.4	1223.6	1.4054
650	0.564	1184.7	1289.1	1.4822	0.450	1170.9	1270.8	1.4490
700	0.608	1212.0	1324.6	1.5135	0.491	1201.3	1310.2	1.4837
800	0.688	1261.2	1388.5	1.5665	0.562	1253.7	1378.4	1.5402
900	0.761	1307.3	1448.1	1.6120	0.626	1301.5	1440.4	1.5876
1000	0.831	1352.2	1505.9	1.6530	0.685	1347.5	1499.7	1.6297
1100	0.898	1396.8	1562.9	1.6908	0.743	1393.0	1557.9	1.6682
1200	0.963	1441.5	1619.7	1.7261	0.798	1438.3	1615.5	1.7040
1300	1.027	1486.5	1676.5	1.7593	0.853	1483.8	1673.1	1.7377
1400	1.091	1531.9	1733.7	1.7909	0.906	1529.6	1730.7	1.7696
1600	1.215	1624.4	1849.3	1.8499	1.011	1622.6	1847.1	1.8290
	$p = 1400$ lbf/in.² ($T_{sat} = 587.2°F$)				$p = 1600$ lbf/in.² ($T_{sat} = 605.1°F$)			
Sat.	0.302	1096.0	1174.1	1.3461	0.255	1087.4	1162.9	1.3258
600	0.318	1110.9	1193.1	1.3641				
650	0.367	1155.5	1250.5	1.4171	0.303	1137.8	1227.4	1.3852
700	0.406	1189.6	1294.8	1.4562	0.342	1177.0	1278.1	1.4299
800	0.471	1245.8	1367.9	1.5168	0.403	1237.7	1357.0	1.4953
900	0.529	1295.6	1432.5	1.5661	0.466	1289.5	1424.4	1.5468
1000	0.582	1342.8	1493.5	1.6094	0.504	1338.0	1487.1	1.5913
1100	0.632	1389.1	1552.8	1.6487	0.549	1385.2	1547.7	1.6315
1200	0.681	1435.1	1611.4	1.6851	0.592	1431.8	1607.1	1.6684
1300	0.728	1481.1	1669.6	1.7192	0.634	1478.3	1666.1	1.7029
1400	0.774	1527.2	1727.8	1.7513	0.675	1524.9	1724.8	1.7354
1600	0.865	1620.8	1844.8	1.8111	0.755	1619.0	1842.6	1.7955
	$p = 1800$ lbf/in.² ($T_{sat} = 621.2°F$)				$p = 2000$ lbf/in.² ($T_{sat} = 636.0°F$)			
Sat.	0.218	1077.7	1150.4	1.3060	0.188	1066.6	1136.3	1.2861
650	0.251	1117.0	1200.4	1.3517	0.206	1091.1	1167.2	1.3141
700	0.291	1163.1	1259.9	1.4042	0.249	1147.7	1239.8	1.3782
750	0.322	1198.6	1305.9	1.4430	0.280	1187.3	1291.1	1.4216
800	0.350	1229.1	1345.7	1.4753	0.307	1220.1	1333.8	1.4562
900	0.399	1283.2	1416.1	1.5291	0.353	1276.8	1407.6	1.5126
1000	0.443	1333.1	1480.7	1.5749	0.395	1328.1	1474.1	1.5598
1100	0.484	1381.2	1542.5	1.6159	0.433	1377.2	1537.2	1.6017
1200	0.524	1428.5	1602.9	1.6534	0.469	1425.2	1598.6	1.6398
1300	0.561	1475.5	1662.5	1.6883	0.503	1472.7	1659.0	1.6751
1400	0.598	1522.5	1721.8	1.7211	0.537	1520.2	1718.8	1.7082
1600	0.670	1617.2	1840.4	1.7817	0.602	1615.4	1838.2	1.7692

TABLE A-4E (*Continued*)

T °F	v ft³/lb	u Btu/lb	h Btu/lb	s Btu/lb · °R	v ft³/lb	u Btu/lb	h Btu/lb	s Btu/lb · °R
	$p = 2500$ lbf/in.² ($T_{sat} = 668.3°F$)				$p = 3000$ lbf/in.² ($T_{sat} = 695.5°F$)			
Sat.	0.1306	1031.0	1091.4	1.2327	0.0840	968.8	1015.5	1.1575
700	0.1684	1098.7	1176.6	1.3073	0.0977	1003.9	1058.1	1.1944
750	0.2030	1155.2	1249.1	1.3686	0.1483	1114.7	1197.1	1.3122
800	0.2291	1195.7	1301.7	1.4112	0.1757	1167.6	1265.2	1.3675
900	0.2712	1259.9	1385.4	1.4752	0.2160	1241.8	1361.7	1.4414
1000	0.3069	1315.2	1457.2	1.5262	0.2485	1301.7	1439.6	1.4967
1100	0.3393	1366.8	1523.8	1.5704	0.2772	1356.2	1510.1	1.5434
1200	0.3696	1416.7	1587.7	1.6101	0.3086	1408.0	1576.6	1.5848
1300	0.3984	1465.7	1650.0	1.6465	0.3285	1458.5	1640.9	1.6224
1400	0.4261	1514.2	1711.3	1.6804	0.3524	1508.1	1703.7	1.6571
1500	0.4531	1562.5	1772.1	1.7123	0.3754	1557.3	1765.7	1.6896
1600	0.4795	1610.8	1832.6	1.7424	0.3978	1606.3	1827.1	1.7201
	$p = 3500$ lbf/in.²				$p = 4000$ lbf/in.²			
650	0.0249	663.5	679.7	0.8630	0.0245	657.7	675.8	0.8574
700	0.0306	759.5	779.3	0.9506	0.0287	742.1	763.4	0.9345
750	0.1046	1058.4	1126.1	1.2440	0.0633	960.7	1007.5	1.1395
800	0.1363	1134.7	1223.0	1.3226	0.1052	1095.0	1172.9	1.2740
900	0.1763	1222.4	1336.5	1.4096	0.1462	1201.5	1309.7	1.3789
1000	0.2066	1287.6	1421.4	1.4699	0.1752	1272.9	1402.6	1.4449
1100	0.2328	1345.2	1496.0	1.5193	0.1995	1333.9	1481.6	1.4973
1200	0.2566	1399.2	1565.3	1.5624	0.2213	1390.1	1553.9	1.5423
1300	0.2787	1451.1	1631.7	1.6012	0.2414	1443.7	1622.4	1.5823
1400	0.2997	1501.9	1696.1	1.6368	0.2603	1495.7	1688.4	1.6188
1500	0.3199	1552.0	1759.2	1.6699	0.2784	1546.7	1752.8	1.6526
1600	0.3395	1601.7	1831.6	1.7010	0.2959	1597.1	1816.1	1.6841
	$p = 4400$ lbf/in.²				$p = 4800$ lbf/in.²			
650	0.0242	653.6	673.3	0.8535	0.0237	649.8	671.0	0.8499
700	0.0278	732.7	755.3	0.9257	0.0271	725.1	749.1	0.9187
750	0.0415	870.8	904.6	1.0513	0.0352	832.6	863.9	1.0154
800	0.0844	1056.5	1125.3	1.2306	0.0668	1011.2	1070.5	1.1827
900	0.1270	1183.7	1287.1	1.3548	0.1109	1164.8	1263.4	1.3310
1000	0.1552	1260.8	1387.2	1.4260	0.1385	1248.3	1317.4	1.4078
1100	0.1784	1324.7	1469.9	1.4809	0.1608	1315.3	1458.1	1.4653
1200	0.1989	1382.8	1544.7	1.5274	0.1802	1375.4	1535.4	1.5133
1300	0.2176	1437.7	1614.9	1.5685	0.1979	1431.7	1607.4	1.5555
1400	0.2352	1490.7	1682.3	1.6057	0.2143	1485.7	1676.1	1.5934
1500	0.2520	1542.7	1747.6	1.6399	0.2300	1538.2	1742.5	1.6282
1600	0.2681	1593.4	1811.7	1.6718	0.2450	1589.8	1807.4	1.6605

TABLE A-5E Properties of Compressed Liquid Water

T	v	u	h	s	v	u	h	s
°F	ft³/lb	Btu/lb	Btu/lb	Btu/lb · °R	ft³/lb	Btu/lb	Btu/lb	Btu/lb · °R
	$p = 500$ lbf/in.² ($T_{sat} = 467.1$°F)				$p = 1000$ lbf/in.² ($T_{sat} = 544.7$°F)			
32	0.015994	0.00	1.49	0.00000	0.015967	0.03	2.99	0.00005
50	0.015998	18.02	19.50	0.03599	0.015972	17.99	20.94	0.03592
100	0.016106	67.87	69.36	0.12932	0.016082	67.70	70.68	0.12901
150	0.016318	117.66	119.17	0.21457	0.016293	117.38	120.40	0.21410
200	0.016608	167.65	169.19	0.29341	0.016580	167.26	170.32	0.29281
300	0.017416	268.92	270.53	0.43641	0.017379	268.24	271.46	0.43552
400	0.018608	373.68	375.40	0.56604	0.018550	372.55	375.98	0.56472
Sat.	0.019748	447.70	449.53	0.64904	0.021591	538.39	542.38	0.74320
	$p = 1500$ lbf/in.² ($T_{sat} = 596.4$°F)				$p = 2000$ lbf/in.² ($T_{sat} = 636.0$°F)			
32	0.015939	0.05	4.47	0.00007	0.015912	0.06	5.95	0.00008
50	0.015946	17.95	22.38	0.03584	0.015920	17.91	23.81	0.03575
100	0.016058	67.53	71.99	0.12870	0.016034	67.37	73.30	0.12839
150	0.016268	117.10	121.62	0.21364	0.016244	116.83	122.84	0.21318
200	0.016554	166.87	171.46	0.29221	0.016527	166.49	172.60	0.29162
300	0.017343	267.58	272.39	0.43463	0.017308	266.93	273.33	0.43376
400	0.018493	371.45	376.59	0.56343	0.018439	370.38	377.21	0.56216
500	0.02024	481.8	487.4	0.6853	0.02014	479.8	487.3	0.6832
Sat.	0.02346	605.0	611.5	0.8082	0.02565	662.4	671.9	0.8623
	$p = 3000$ lbf/in.² ($T_{sat} = 695.5$°F)				$p = 4000$ lbf/in.²			
32	0.015859	0.09	8.90	0.00009	0.015807	0.10	11.80	0.00005
50	0.015870	17.84	26.65	0.03555	0.015821	17.76	29.47	0.03534
100	0.015987	67.04	75.91	0.12777	0.015942	66.72	78.52	0.12714
150	0.016196	116.30	125.29	0.21226	0.016150	115.77	127.73	0.21136
200	0.016476	165.74	174.89	0.29046	0.016425	165.02	177.18	0.28931
300	0.017240	265.66	275.23	0.43205	0.017174	264.43	277.15	0.43038
400	0.018334	368.32	378.50	0.55970	0.018235	366.35	379.85	0.55734
500	0.019944	476.2	487.3	0.6794	0.019766	472.9	487.5	0.6758
Sat.	0.034310	783.5	802.5	0.9732				

TABLE A-6E Properties of Saturated Water (Solid–Vapor): Temperature Table

Temp. °F	Press. lbf/in²	Specific volume ft³/lb		Internal energy Btu/lb			Enthalpy Btu/lb			Entropy Btu/lb·°R		
		Sat. Solid v_i	Sat. Vapor $v_g \times 10^{-3}$	Sat. Solid u_i	Subl. u_{ig}	Sat. Vapor u_g	Sat. Solid h_i	Subl. h_{ig}	Sat. Vapor h_g	Sat. Solid s_i	Subl. s_{ig}	Sat. Vapor s_g
32.018	.0887	.01747	3.302	−143.34	1164.6	1021.2	−143.34	1218.7	1075.4	−.292	2.479	2.187
32	.0886	.01747	3.305	−143.35	1164.6	1021.2	−143.35	1218.7	1075.4	−.292	2.479	2.187
30	.0808	.01747	3.607	−144.35	1164.9	1020.5	−144.35	1218.9	1074.5	−.294	2.489	2.195
25	.0641	.01746	4.506	−146.84	1165.7	1018.9	−146.84	1219.1	1072.3	−.299	2.515	2.216
20	.0505	.01745	5.655	−149.31	1166.5	1017.2	−149.31	1219.4	1070.1	−.304	2.542	2.238
15	.0396	.01745	7.13	−151.75	1167.3	1015.5	−151.75	1219.7	1067.9	−.309	2.569	2.260
10	.0309	.01744	9.04	−154.17	1168.1	1013.9	−154.17	1219.9	1065.7	−.314	2.597	2.283
5	.0240	.01743	11.52	−156.56	1168.8	1012.2	−156.56	1220.1	1063.5	−.320	2.626	2.306
0	.0185	.01743	14.77	−158.93	1169.5	1010.6	−158.93	1220.2	1061.2	−.325	2.655	2.330
−5	.0142	.01742	19.03	−161.27	1170.2	1008.9	−161.27	1220.3	1059.0	−.330	2.684	2.354
−10	.0109	.01741	24.66	−163.59	1170.9	1007.3	−163.59	1220.4	1056.8	−.335	2.714	2.379
−15	.0082	.01740	32.2	−165.89	1171.5	1005.6	−165.89	1220.5	1054.6	−.340	2.745	2.405
−20	.0062	.01740	42.2	−168.16	1172.1	1003.9	−168.16	1220.6	1052.4	−.345	2.776	2.431
−25	.0046	.01739	55.7	−170.40	1172.7	1002.3	−170.40	1220.6	1050.2	−.351	2.808	2.457
−30	.0035	.01738	74.1	−172.63	1173.2	1000.6	−172.63	1220.6	1048.0	−.356	2.841	2.485
−35	.0026	.01737	99.2	−174.82	1173.8	998.9	−174.82	1220.6	1045.8	−.361	2.874	2.513
−40	.0019	.01737	133.8	−177.00	1174.3	997.3	−177.00	1220.6	1043.6	−.366	2.908	2.542

Source: J. H. Keenan, F. G. Keyes, P. G. Hill, and J. G. Moore, *Steam Tables,* Wiley, New York, 1969.

TABLE A-7E Properties of Saturated Refrigerant 12 (Liquid–Vapor): Temperature Table

Temp. °F	Press. lbf/in.²	Specific volume ft³/lb		Internal energy Btu/lb		Enthalpy Btu/lb			Entropy Btu/lb · °R		Temp. °F
		Sat. liquid v_f	Sat. vapor v_g	Sat. liquid u_f	Sat. vapor u_g	Sat. liquid h_f	Evap. h_{fg}	Sat. vapor h_g	Sat. liquid s_f	Sat. vapor s_g	
−40	9.308	0.01056	3.8750	−0.02	66.24	0	72.91	72.91	0	0.1737	−40
−30	11.999	0.01067	3.0585	1.93	67.22	2.11	71.90	74.01	0.0050	0.1723	−30
−20	15.267	0.01079	2.4429	4.21	68.21	4.24	70.87	75.11	0.0098	0.1710	−20
−15	17.141	0.01085	2.1924	5.27	68.69	5.30	70.35	75.65	0.0122	0.1704	−15
−10	19.189	0.01091	1.9727	6.33	69.19	6.37	69.82	76.19	0.0146	0.1699	−10
−5	21.422	0.01097	1.7794	7.40	69.67	7.44	69.29	76.73	0.0170	0.1694	−5
0	23.849	0.01103	1.6089	8.47	70.17	8.52	68.75	77.27	0.0193	0.1689	0
5	26.483	0.01109	1.4580	9.55	70.65	9.60	68.20	77.80	0.0216	0.1684	5
10	29.335	0.01116	1.3241	10.62	71.15	10.68	67.65	78.33	0.0240	0.1680	10
20	35.736	0.01130	1.0988	12.79	72.12	12.86	66.52	79.38	0.0285	0.1672	20
25	39.310	0.01137	1.0039	13.88	72.61	13.96	65.95	79.91	0.0308	0.1668	25
30	43.148	0.01144	0.9188	14.97	73.08	15.06	65.36	80.42	0.0330	0.1665	30
40	51.667	0.01159	0.7736	17.16	74.04	17.27	64.16	81.43	0.0375	0.1659	40
50	61.394	0.01175	0.6554	19.38	74.99	19.51	62.93	82.44	0.0418	0.1653	50
60	72.433	0.01191	0.5584	21.61	75.92	21.77	61.64	83.41	0.0462	0.1648	60
70	84.89	0.01209	0.4782	23.86	76.85	24.05	60.31	84.36	0.0505	0.1643	70
80	98.87	0.01228	0.4114	26.14	77.76	26.37	58.92	85.29	0.0548	0.1639	80
85	106.47	0.01238	0.3821	27.29	78.20	27.53	58.20	85.73	0.0569	0.1637	85
90	114.49	0.01248	0.3553	28.45	78.65	28.71	57.46	86.17	0.0590	0.1635	90
95	122.95	0.01258	0.3306	29.61	79.09	29.90	56.71	86.61	0.0611	0.1633	95
100	131.86	0.01269	0.3079	30.79	79.51	31.10	55.93	87.03	0.0632	0.1632	100
105	141.25	0.01281	0.2870	31.98	79.94	32.31	55.13	87.44	0.0653	0.1630	105
110	151.11	0.01292	0.2677	33.16	80.36	33.53	54.31	87.84	0.0675	0.1628	110
115	161.47	0.01305	0.2498	34.37	80.76	34.76	53.47	88.23	0.0696	0.1626	115
120	172.35	0.01317	0.2333	35.59	81.17	36.01	52.60	88.61	0.0717	0.1624	120
140	221.32	0.01375	0.1780	40.60	82.68	41.16	48.81	89.97	0.0802	0.1616	140
160	279.82	0.01445	0.1360	45.88	83.96	46.63	44.37	91.00	0.0889	0.1605	160
180	349.00	0.01536	0.1033	51.57	84.89	52.56	39.00	91.56	0.0980	0.1590	180
200	430.09	0.01666	0.0767	57.87	85.17	59.20	32.08	91.28	0.1079	0.1565	200
233.6	596.9	0.0287	0.0287	75.69	75.69	78.86	0	78.86	0.1359	0.1359	233.6

Source: Tables A-7E through A-9E are adapted from K. Wark, *Thermodynamics, 4th ed.,* McGraw-Hill, New York, 1983, as based on data from Freon Products Division, E. I. du Pont de Nemours & Company, 1956.

TABLE A-8E Properties of Saturated Refrigerant 12 (Liquid–Vapor): Pressure Table

Press. lbf/in.²	Temp. °F	Specific volume ft³/lb		Internal energy Btu/lb		Enthalpy Btu/lb			Entropy Btu/lb · °R		Press. lbf/in.²
		Sat. liquid v_f	Sat. vapor v_g	Sat. liquid u_f	Sat. vapor u_g	Sat. liquid h_f	Evap. h_{fg}	Sat. vapor h_g	Sat. liquid s_f	Sat. vapor s_g	
5	−62.35	0.0103	6.9069	−4.69	64.04	−4.68	75.11	70.43	−0.0114	0.1776	5
10	−37.23	0.0106	3.6246	0.56	66.51	0.58	72.64	73.22	0.0014	0.1733	10
15	−20.75	0.0108	2.4835	4.05	68.13	4.08	70.95	75.03	0.0095	0.1711	15
20	−8.13	0.0109	1.8977	6.73	69.37	6.77	69.63	76.40	0.0155	0.1697	20
30	11.11	0.0112	1.2964	10.86	71.25	10.93	67.53	78.45	0.0245	0.1679	30
40	25.93	0.0114	0.9874	14.08	72.69	14.16	65.84	80.00	0.0312	0.1668	40
50	38.15	0.0116	0.7982	16.75	73.86	16.86	64.39	81.25	0.0366	0.1660	50
60	48.64	0.0117	0.6701	19.07	74.86	19.20	63.10	82.30	0.0413	0.1654	60
70	57.90	0.0119	0.5772	21.13	75.73	21.29	61.92	83.21	0.0453	0.1649	70
80	66.21	0.0120	0.5068	23.00	76.50	23.18	60.82	84.00	0.0489	0.1645	80
90	73.79	0.0122	0.4514	24.72	77.20	24.92	59.79	84.71	0.0521	0.1642	90
100	80.76	0.0123	0.4067	26.31	77.82	26.54	58.81	85.35	0.0551	0.1639	100
120	93.29	0.0126	0.3389	29.21	78.93	29.49	56.97	86.46	0.0604	0.1634	120
140	104.35	0.0128	0.2896	31.82	79.89	32.15	55.24	87.39	0.0651	0.1630	140
160	114.30	0.0130	0.2522	34.21	80.71	34.59	53.59	88.18	0.0693	0.1626	160
180	123.38	0.0133	0.2228	36.42	81.44	36.86	52.00	88.86	0.0731	0.1623	180
200	131.74	0.0135	0.1989	38.50	82.08	39.00	50.44	89.44	0.0767	0.1620	200
220	139.51	0.0137	0.1792	40.48	82.08	41.03	48.90	89.94	0.0816	0.1616	220
240	146.77	0.0140	0.1625	42.35	83.14	42.97	47.39	90.36	0.0831	0.1613	240
260	153.60	0.0142	0.1483	44.16	83.58	44.84	45.88	90.72	0.0861	0.1609	260
280	160.06	0.0145	0.1359	45.90	83.97	46.65	44.36	91.01	0.0890	0.1605	280
300	166.18	0.0147	0.1251	47.59	84.30	48.41	42.83	91.24	0.0917	0.1601	300

TABLE A-9E Properties of Superheated Refrigerant 12

T °F	v ft³/lb	u Btu/lb	h Btu/lb	s Btu/lb · °R	v ft³/lb	u Btu/lb	h Btu/lb	s Btu/lb · °R
	p = 10 lbf/in.² (T_sat = −37.23°F)				p = 15 lbf/in.² (T_sat = −20.75°F)			
Sat.	3.6246	66.512	73.219	0.1733	2.4835	68.134	75.028	0.1711
0	3.9809	70.879	78.246	0.1847	2.6201	70.629	77.902	0.1775
20	4.1691	73.299	81.014	0.1906	2.7494	73.080	80.712	0.1835
40	4.3556	75.768	83.828	0.1964	2.8770	75.575	83.561	0.1893
60	4.5408	78.286	86.689	0.2020	3.0031	78.115	86.451	0.1950
80	4.7248	80.853	89.596	0.2075	3.1281	80.700	89.383	0.2005
100	4.9079	83.466	92.548	0.2128	3.2521	83.330	92.357	0.2059
120	5.0903	86.126	95.546	0.2181	3.3754	86.004	95.373	0.2112
140	5.2720	88.830	98.586	0.2233	3.4981	88.719	98.429	0.2164
160	5.4533	91.578	101.669	0.2283	3.6202	91.476	101.525	0.2215
180	5.6341	94.367	104.793	0.2333	3.7419	94.274	104.661	0.2265
200	5.8145	97.197	107.957	0.2381	3.8632	97.112	107.835	0.2314

T °F	v ft³/lb	u Btu/lb	h Btu/lb	s Btu/lb · °R	v ft³/lb	u Btu/lb	h Btu/lb	s Btu/lb · °R
	p = 20 lbf/in.² (T_sat = −8.13°F)				p = 30 lbf/in.² (T_sat = 11.11°F)			
Sat.	1.8977	69.374	76.397	0.1697	1.2964	71.255	78.452	0.1679
20	2.0391	72.856	80.403	0.1783	1.3278	72.394	79.765	0.1707
40	2.1373	75.379	83.289	0.1842	1.3969	74.975	82.730	0.1767
60	2.2340	77.942	86.210	0.1899	1.4644	77.586	85.716	0.1826
80	2.3295	80.546	89.168	0.1955	1.5306	80.232	88.729	0.1883
100	2.4241	83.192	92.164	0.2010	1.5957	82.911	91.770	0.1938
120	2.5179	85.879	95.198	0.2063	1.6600	85.627	94.843	0.1992
140	2.6110	88.607	98.270	0.2115	1.7237	88.379	97.948	0.2045
160	2.7036	91.374	101.380	0.2166	1.7868	91.166	101.086	0.2096
180	2.7957	94.181	104.528	0.2216	1.8494	93.991	104.258	0.2146
200	2.8874	97.026	107.712	0.2265	1.9116	96.852	107.464	0.2196
220	2.9789	99.907	110.932	0.2313	1.9735	99.746	110.702	0.2244

T °F	v ft³/lb	u Btu/lb	h Btu/lb	s Btu/lb · °R	v ft³/lb	u Btu/lb	h Btu/lb	s Btu/lb · °R
	p = 40 lbf/in.² (T_sat = 25.93°F)				p = 50 lbf/in.² (T_sat = 38.15°F)			
Sat.	0.9874	72.691	80.000	0.1668	0.7982	73.863	81.249	0.1660
40	1.0258	74.555	82.148	0.1711	0.8025	74.115	81.540	0.1666
60	1.0789	77.220	85.206	0.1771	0.8471	76.838	84.676	0.1727
80	1.1306	79.908	88.277	0.1829	0.8903	79.574	87.811	0.1786
100	1.1812	82.624	91.367	0.1885	0.9322	82.328	90.953	0.1843
120	1.2309	85.369	94.480	0.1940	0.9731	85.106	94.110	0.1899
140	1.2798	88.147	97.620	0.1993	1.0133	87.910	97.286	0.1953
160	1.3282	90.957	100.788	0.2045	1.0529	90.743	100.485	0.2005
180	1.3761	93.800	103.985	0.2096	1.0920	93.604	103.708	0.2056
200	1.4236	96.674	107.212	0.2146	1.1307	96.496	106.958	0.2106
220	1.4707	99.583	110.469	0.2194	1.1690	99.419	110.235	0.2155
240	1.5176	102.524	113.757	0.2242	1.2070	102.371	113.539	0.2203

TABLE A-9E (*Continued*)

T °F	v ft³/lb	u Btu/lb	h Btu/lb	s Btu/lb·°R	v ft³/lb	u Btu/lb	h Btu/lb	s Btu/lb·°R
	$p = 60$ lbf/in.² ($T_{sat} = 48.64$°F)				$p = 70$ lbf/in.² ($T_{sat} = 57.90$°F)			
Sat.	0.6701	74.859	82.299	0.1654	0.5772	75.729	83.206	0.1649
60	0.6921	76.442	84.126	0.1689	0.5809	76.027	83.552	0.1656
80	0.7296	79.229	87.330	0.1750	0.6146	78.871	86.832	0.1718
100	0.7659	82.024	90.528	0.1808	0.6469	81.712	90.091	0.1777
120	0.8011	84.836	93.731	0.1864	0.6780	84.560	93.343	0.1834
140	0.8355	87.668	96.945	0.1919	0.7084	87.421	96.597	0.1889
160	0.8693	90.524	100.776	0.1972	0.7380	90.302	99.862	0.1943
180	0.9025	93.406	103.427	0.2023	0.7671	93.205	103.141	0.1995
200	0.9353	96.315	106.700	0.2074	0.7957	96.132	106.439	0.2046
220	0.9678	99.252	109.997	0.2123	0.8240	99.083	109.756	0.2095
240	0.9998	102.217	113.319	0.2171	0.8519	102.061	113.096	0.2144
260	1.0318	105.210	116.666	0.2218	0.8796	105.065	116.459	0.2191
	$p = 80$ lbf/in.² ($T_{sat} = 66.21$°F)				$p = 90$ lbf/in.² ($T_{sat} = 73.79$°F)			
Sat.	0.5068	76.500	84.003	0.1645	0.4514	77.194	84.713	0.1642
80	0.5280	78.500	86.316	0.1689	0.4602	78.115	85.779	0.1662
100	0.5573	81.389	89.640	0.1749	0.4875	81.056	89.175	0.1723
120	0.5856	84.276	92.945	0.1807	0.5135	83.984	92.536	0.1782
140	0.6129	87.169	96.242	0.1863	0.5385	86.911	95.879	0.1839
160	0.6394	90.076	99.542	0.1917	0.5627	89.845	99.216	0.1894
180	0.6654	93.000	102.851	0.1970	0.5863	92.793	102.557	0.1947
200	0.6910	95.945	106.174	0.2021	0.6094	95.755	105.905	0.1998
220	0.7161	98.912	109.513	0.2071	0.6321	98.739	109.267	0.2049
240	0.7409	101.904	112.872	0.2119	0.6545	101.743	112.644	0.2098
260	0.7654	104.919	116.251	0.2167	0.6766	104.771	116.040	0.2146
280	0.7898	107.960	119.652	0.2214	0.6985	107.823	119.456	0.2192
	$p = 100$ lbf/in.² ($T_{sat} = 80.76$°F)				$p = 120$ lbf/in.² ($T_{sat} = 93.29$°F)			
Sat.	0.4067	77.824	85.351	0.1639	0.3389	78.933	86.459	0.1634
100	0.4314	80.711	88.694	0.1700	0.3466	79.978	87.675	0.1656
120	0.4556	83.685	92.116	0.1760	0.3684	83.056	91.237	0.1718
140	0.4788	86.647	95.507	0.1817	0.3890	86.098	94.736	0.1778
160	0.5012	89.610	98.884	0.1873	0.4087	89.123	98.199	0.1835
180	0.5229	92.580	102.257	0.1926	0.4277	92.144	101.642	0.1889
200	0.5441	95.564	105.633	0.1978	0.4461	95.170	105.076	0.1942
220	0.5649	98.564	109.018	0.2029	0.4640	98.205	108.509	0.1993
240	0.5854	101.582	112.415	0.2078	0.4816	101.253	111.948	0.2043
260	0.6055	104.622	115.828	0.2126	0.4989	104.317	115.396	0.2092
280	0.6255	107.684	119.258	0.2173	0.5159	107.401	118.857	0.2139
300	0.6452	110.768	122.707	0.2219	0.5327	110.504	122.333	0.2186

TABLE A-9E (*Continued*)

T	v	u	h	s	v	u	h	s
°F	ft³/lb	Btu/lb	Btu/lb	Btu/lb · °R	ft³/lb	Btu/lb	Btu/lb	Btu/lb · °R

	$p = 140$ lbf/in.² ($T_{sat} = 104.35°F$)				$p = 160$ lbf/in.² ($T_{sat} = 114.30°F$)			
Sat.	0.2896	79.886	87.389	0.1630	0.2522	80.713	88.180	0.1626
120	0.3055	82.382	90.297	0.1681	0.2576	81.656	89.283	0.1645
140	0.3245	85.516	93.923	0.1742	0.2756	84.899	93.059	0.1709
160	0.3423	88.615	97.483	0.1801	0.2922	88.080	96.732	0.1770
180	0.3594	91.692	101.003	0.1857	0.3080	91.221	100.340	0.1827
200	0.3758	94.765	104.501	0.1910	0.3230	94.344	103.907	0.1882
220	0.3918	97.837	107.987	0.1963	0.3375	97.457	107.450	0.1935
240	0.4073	100.918	111.470	0.2013	0.3516	100.570	110.980	0.1986
260	0.4226	104.008	114.956	0.2062	0.3653	103.690	114.506	0.2036
280	0.4375	107.115	118.449	0.2110	0.3787	106.820	118.033	0.2084
300	0.4523	110.235	121.953	0.2157	0.3919	109.964	121.567	0.2131
320	0.4668	113.376	125.470	0.2202	0.4049	113.121	125.109	0.2177

	$p = 180$ lbf/in.² ($T_{sat} = 123.38°F$)				$p = 200$ lbf/in.² ($T_{sat} = 131.74°F$)			
Sat.	0.2228	81.436	88.857	0.1623	0.1989	82.077	89.439	0.1620
140	0.2371	84.238	92.136	0.1678	0.2058	83.521	91.137	0.1648
160	0.2530	87.513	95.940	0.1741	0.2212	86.913	95.100	0.1713
180	0.2678	90.727	99.647	0.1800	0.2354	90.211	98.921	0.1774
200	0.2818	93.904	103.291	0.1856	0.2486	93.451	102.652	0.1831
220	0.2952	97.063	106.896	0.1910	0.2612	96.659	106.325	0.1886
240	0.3081	100.215	110.478	0.1961	0.2732	99.850	109.962	0.1939
260	0.3207	103.364	114.046	0.2012	0.2849	103.032	113.576	0.1990
280	0.3329	106.521	117.610	0.2061	0.2962	106.214	117.178	0.2039
300	0.3449	109.686	121.174	0.2108	0.3073	109.402	120.775	0.2087
320	0.3567	112.863	124.744	0.2155	0.3182	112.598	124.373	0.2134
340	0.3683	116.053	128.321	0.2200	0.3288	115.805	127.974	0.2179

	$p = 300$ lbf/in.² ($T_{sat} = 166.18°F$)				$p = 400$ lbf/in.² ($T_{sat} = 192.93°F$)			
Sat.	0.1251	84.295	91.240	0.1601	0.0856	85.178	91.513	0.1576
180	0.1348	87.071	94.556	0.1654				
200	0.1470	90.816	98.975	0.1722	0.0910	86.982	93.718	0.1609
220	0.1577	94.379	103.136	0.1784	0.1032	91.410	99.046	0.1689
240	0.1676	97.835	107.140	0.1842	0.1130	95.371	103.735	0.1757
260	0.1769	101.225	111.043	0.1897	0.1216	99.102	108.105	0.1818
280	0.1856	104.574	114.879	0.1950	0.1295	102.701	112.286	0.1876
300	0.1940	107.899	118.670	0.2000	0.1368	106.217	116.343	0.1930
320	0.2021	111.208	122.430	0.2049	0.1437	109.680	120.318	0.1981
340	0.2100	114.512	126.171	0.2096	0.1503	113.108	124.235	0.2031
360	0.2177	117.814	129.900	0.2142	0.1567	116.514	128.112	0.2079

TABLE A-10E Properties of Saturated Refrigerant 134a (Liquid–Vapor): Temperature Table

Temp. °F	Press. lbf/in.²	Specific Volume ft³/lb		Internal Energy Btu/lb		Enthalpy Btu/lb			Entropy Btu/lb · °R		Temp. °F
		Sat. Liquid v_f	Sat. Vapor v_g	Sat. Liquid u_f	Sat. Vapor u_g	Sat. Liquid h_f	Evap. h_{fg}	Sat. Vapor h_g	Sat. Liquid s_f	Sat. Vapor s_g	
−40	7.490	0.01130	5.7173	−0.02	87.90	0.00	95.82	95.82	0.0000	0.2283	−40
−30	9.920	0.01143	4.3911	2.81	89.26	2.83	94.49	97.32	0.0067	0.2266	−30
−20	12.949	0.01156	3.4173	5.69	90.62	5.71	93.10	98.81	0.0133	0.2250	−20
−15	14.718	0.01163	3.0286	7.14	91.30	7.17	92.38	99.55	0.0166	0.2243	−15
−10	16.674	0.01170	2.6918	8.61	91.98	8.65	91.64	100.29	0.0199	0.2236	−10
−5	18.831	0.01178	2.3992	10.09	92.66	10.13	90.89	101.02	0.0231	0.2230	−5
0	21.203	0.01185	2.1440	11.58	93.33	11.63	90.12	101.75	0.0264	0.2224	0
5	23.805	0.01193	1.9208	13.09	94.01	13.14	89.33	102.47	0.0296	0.2219	5
10	26.651	0.01200	1.7251	14.60	94.68	14.66	88.53	103.19	0.0329	0.2214	10
15	29.756	0.01208	1.5529	16.13	95.35	16.20	87.71	103.90	0.0361	0.2209	15
20	33.137	0.01216	1.4009	17.67	96.02	17.74	86.87	104.61	0.0393	0.2205	20
25	36.809	0.01225	1.2666	19.22	96.69	19.30	86.02	105.32	0.0426	0.2200	25
30	40.788	0.01233	1.1474	20.78	97.35	20.87	85.14	106.01	0.0458	0.2196	30
40	49.738	0.01251	0.9470	23.94	98.67	24.05	83.34	107.39	0.0522	0.2189	40
50	60.125	0.01270	0.7871	27.14	99.98	27.28	81.46	108.74	0.0585	0.2183	50
60	72.092	0.01290	0.6584	30.39	101.27	30.56	79.49	110.05	0.0648	0.2178	60
70	85.788	0.01311	0.5538	33.68	102.54	33.89	77.44	111.33	0.0711	0.2173	70
80	101.37	0.01334	0.4682	37.02	103.78	37.27	75.29	112.56	0.0774	0.2169	80
85	109.92	0.01346	0.4312	38.72	104.39	38.99	74.17	113.16	0.0805	0.2167	85
90	118.99	0.01358	0.3975	40.42	105.00	40.72	73.03	113.75	0.0836	0.2165	90
95	128.62	0.01371	0.3668	42.14	105.60	42.47	71.86	114.33	0.0867	0.2163	95
100	138.83	0.01385	0.3388	43.87	106.18	44.23	70.66	114.89	0.0898	0.2161	100
105	149.63	0.01399	0.3131	45.62	106.76	46.01	69.42	115.43	0.0930	0.2159	105
110	161.04	0.01414	0.2896	47.39	107.33	47.81	68.15	115.96	0.0961	0.2157	110
115	173.10	0.01429	0.2680	49.17	107.88	49.63	66.84	116.47	0.0992	0.2155	115
120	185.82	0.01445	0.2481	50.97	108.42	51.47	65.48	116.95	0.1023	0.2153	120
140	243.86	0.01520	0.1827	58.39	110.41	59.08	59.57	118.65	0.1150	0.2143	140
160	314.63	0.01617	0.1341	66.26	111.97	67.20	52.58	119.78	0.1280	0.2128	160
180	400.22	0.01758	0.0964	74.83	112.77	76.13	43.78	119.91	0.1417	0.2101	180
200	503.52	0.02014	0.0647	84.90	111.66	86.77	30.92	117.69	0.1575	0.2044	200
210	563.51	0.02329	0.0476	91.84	108.48	94.27	19.18	113.45	0.1684	0.1971	210

Source: Tables A-10E through A-12E are calculated based on equations from D. P. Wilson and R. S. Basu, "Thermodynamic Properties of a New Stratospherically Safe Working Fluid—Refrigerant 134a," *ASHRAE Trans.,* Vol. 94, Pt. 2, 1988, pp. 2095–2118.

TABLE A-11E Properties of Saturated Refrigerant 134a (Liquid–Vapor): Pressure Table

Press. lbf/in.^{2}	Temp. °F	Specific Volume ft^3/lb		Internal Energy Btu/lb		Enthalpy Btu/lb			Entropy Btu/lb · °R		Press. lbf/in.^{2}
		Sat. Liquid v_f	Sat. Vapor v_g	Sat. Liquid u_f	Sat. Vapor u_g	Sat. Liquid h_f	Evap. h_{fg}	Sat. Vapor h_g	Sat. Liquid s_f	Sat. Vapor s_g	
5	−53.48	0.01113	8.3508	−3.74	86.07	−3.73	97.53	93.79	−.0090	0.2311	5
10	−29.71	0.01143	4.3581	2.89	89.30	2.91	94.45	97.37	0.0068	0.2265	10
15	−14.25	0.01164	2.9747	7.36	91.40	7.40	92.27	99.66	0.0171	0.2242	15
20	−2.48	0.01181	2.2661	10.84	93.00	10.89	90.50	101.39	0.0248	0.2227	20
30	15.38	0.01209	1.5408	16.24	95.40	16.31	87.65	103.96	0.0364	0.2209	30
40	29.04	0.01232	1.1692	20.48	97.23	20.57	85.31	105.88	0.0452	0.2197	40
50	40.27	0.01252	0.9422	24.02	98.71	24.14	83.29	107.43	0.0523	0.2189	50
60	49.89	0.01270	0.7887	27.10	99.96	27.24	81.48	108.72	0.0584	0.2183	60
70	58.35	0.01286	0.6778	29.85	101.05	30.01	79.82	109.83	0.0638	0.2179	70
80	65.93	0.01302	0.5938	32.33	102.02	32.53	78.28	110.81	0.0686	0.2175	80
90	72.83	0.01317	0.5278	34.62	102.89	34.84	76.84	111.68	0.0729	0.2172	90
100	79.17	0.01332	0.4747	36.75	103.68	36.99	75.47	112.46	0.0768	0.2169	100
120	90.54	0.01360	0.3941	40.61	105.06	40.91	72.91	113.82	0.0839	0.2165	120
140	100.56	0.01386	0.3358	44.07	106.25	44.43	70.52	114.95	0.0902	0.2161	140
160	109.56	0.01412	0.2916	47.23	107.28	47.65	68.26	115.91	0.0958	0.2157	160
180	117.74	0.01438	0.2569	50.16	108.18	50.64	66.10	116.74	0.1009	0.2154	180
200	125.28	0.01463	0.2288	52.90	108.98	53.44	64.01	117.44	0.1057	0.2151	200
220	132.27	0.01489	0.2056	55.48	109.68	56.09	61.96	118.05	0.1101	0.2147	220
240	138.79	0.01515	0.1861	57.93	110.30	58.61	59.96	118.56	0.1142	0.2144	240
260	144.92	0.01541	0.1695	60.28	110.84	61.02	57.97	118.99	0.1181	0.2140	260
280	150.70	0.01568	0.1550	62.53	111.31	63.34	56.00	119.35	0.1219	0.2136	280
300	156.17	0.01596	0.1424	64.71	111.72	65.59	54.03	119.62	0.1254	0.2132	300
350	168.72	0.01671	0.1166	69.88	112.45	70.97	49.03	120.00	0.1338	0.2118	350
400	179.95	0.01758	0.0965	74.81	112.77	76.11	43.80	119.91	0.1417	0.2102	400
450	190.12	0.01863	0.0800	79.63	112.60	81.18	38.08	119.26	0.1493	0.2079	450
500	199.38	0.02002	0.0657	84.54	111.76	86.39	31.44	117.83	0.1570	0.2047	500

TABLE A-12E Properties of Superheated Refrigerant 134a Vapor

T °F	v ft³/lb	u Btu/lb	h Btu/lb	s Btu/lb · °R	v ft³/lb	u Btu/lb	h Btu/lb	s Btu/lb · °R
	$p = 10$ lbf/in.² ($T_{sat} = -29.71°F$)				$p = 15$ lbf/in.² ($T_{sat} = -14.25°F$)			
Sat.	4.3581	89.30	97.37	0.2265	2.9747	91.40	99.66	0.2242
−20	4.4718	90.89	99.17	0.2307				
0	4.7026	94.24	102.94	0.2391	3.0893	93.84	102.42	0.2303
20	4.9297	97.67	106.79	0.2472	3.2468	97.33	106.34	0.2386
40	5.1539	101.19	110.72	0.2553	3.4012	100.89	110.33	0.2468
60	5.3758	104.80	114.74	0.2632	3.5533	104.54	114.40	0.2548
80	5.5959	108.50	118.85	0.2709	3.7034	108.28	118.56	0.2626
100	5.8145	112.29	123.05	0.2786	3.8520	112.10	122.79	0.2703
120	6.0318	116.18	127.34	0.2861	3.9993	116.01	127.11	0.2779
140	6.2482	120.16	131.72	0.2935	4.1456	120.00	131.51	0.2854
160	6.4638	124.23	136.19	0.3009	4.2911	124.09	136.00	0.2927
180	6.6786	128.38	140.74	0.3081	4.4359	128.26	140.57	0.3000
200	6.8929	132.63	145.39	0.3152	4.5801	132.52	145.23	0.3072
	$p = 20$ lbf/in.² ($T_{sat} = -2.48°F$)				$p = 30$ lbf/in.² ($T_{sat} = 15.38°F$)			
Sat.	2.2661	93.00	101.39	0.2227	1.5408	95.40	103.96	0.2209
0	2.2816	93.43	101.88	0.2238				
20	2.4046	96.98	105.88	0.2323	1.5611	96.26	104.92	0.2229
40	2.5244	100.59	109.94	0.2406	1.6465	99.98	109.12	0.2315
60	2.6416	104.28	114.06	0.2487	1.7293	103.75	113.35	0.2398
80	2.7569	108.05	118.25	0.2566	1.8098	107.59	117.63	0.2478
100	2.8705	111.90	122.52	0.2644	1.8887	111.49	121.98	0.2558
120	2.9829	115.83	126.87	0.2720	1.9662	115.47	126.39	0.2635
140	3.0942	119.85	131.30	0.2795	2.0426	119.53	130.87	0.2711
160	3.2047	123.95	135.81	0.2869	2.1181	123.66	135.42	0.2786
180	3.3144	128.13	140.40	0.2922	2.1929	127.88	140.05	0.2859
200	3.4236	132.40	145.07	0.3014	2.2671	132.17	144.76	0.2932
220	3.5323	136.76	149.83	0.3085	2.3407	136.55	149.54	0.3003
	$p = 40$ lbf/in.² ($T_{sat} = 29.04°F$)				$p = 50$ lbf/in.² ($T_{sat} = 40.27$)			
Sat.	1.1692	97.23	105.88	0.2197	0.9422	98.71	107.43	0.2189
40	1.2065	99.33	108.26	0.2245				
60	1.2723	103.20	112.62	0.2331	0.9974	102.62	111.85	0.2276
80	1.3357	107.11	117.00	0.2414	1.0508	106.62	116.34	0.2361
100	1.3973	111.08	121.42	0.2494	1.1022	110.65	120.85	0.2443
120	1.4575	115.11	125.90	0.2573	1.1520	114.74	125.39	0.2523
140	1.5165	119.21	130.43	0.2650	1.2007	118.88	129.99	0.2601
160	1.5746	123.38	135.03	0.2725	1.2484	123.08	134.64	0.2677
180	1.6319	127.62	139.70	0.2799	1.2953	127.36	139.34	0.2752
200	1.6887	131.94	144.44	0.2872	1.3415	131.71	144.12	0.2825
220	1.7449	136.34	149.25	0.2944	1.3873	136.12	148.96	0.2897
240	1.8006	140.81	154.14	0.3015	1.4326	140.61	153.87	0.2969
260	1.8561	145.36	159.10	0.3085	1.4775	145.18	158.85	0.3039
280	1.9112	149.98	164.13	0.3154	1.5221	149.82	163.90	0.3108

TABLE A-12E (*Continued*)

T °F	v ft³/lb	u Btu/lb	h Btu/lb	s Btu/lb · °R	v ft³/lb	u Btu/lb	h Btu/lb	s Btu/lb · °R
	$p = 60$ lbf/in.² ($T_{sat} = 49.89°F$)				$p = 70$ lbf/in.² ($T_{sat} = 58.35°F$)			
Sat.	0.7887	99.96	108.72	0.2183	0.6778	101.05	109.83	0.2179
60	0.8135	102.03	111.06	0.2229	0.6814	101.40	110.23	0.2186
80	0.8604	106.11	115.66	0.2316	0.7239	105.58	114.96	0.2276
100	0.9051	110.21	120.26	0.2399	0.7640	109.76	119.66	0.2361
120	0.9482	114.35	124.88	0.2480	0.8023	113.96	124.36	0.2444
140	0.9900	118.54	129.53	0.2559	0.8393	118.20	129.07	0.2524
160	1.0308	122.79	134.23	0.2636	0.8752	122.49	133.82	0.2601
180	1.0707	127.10	138.98	0.2712	0.9103	126.83	138.62	0.2678
200	1.1100	131.47	143.79	0.2786	0.9446	131.23	143.46	0.2752
220	1.1488	135.91	148.66	0.2859	0.9784	135.69	148.36	0.2825
240	1.1871	140.42	153.60	0.2930	1.0118	140.22	153.33	0.2897
260	1.2251	145.00	158.60	0.3001	1.0448	144.82	158.35	0.2968
280	1.2627	149.65	163.67	0.3070	1.0774	149.48	163.44	0.3038
300	1.3001	154.38	168.81	0.3139	1.1098	154.22	168.60	0.3107
	$p = 80$ lbf/in.² ($T_{sat} = 65.93°F$)				$p = 90$ lbf/in.² ($T_{sat} = 72.83°F$)			
Sat.	0.5938	102.02	110.81	0.2175	0.5278	102.89	111.68	0.2172
80	0.6211	105.03	114.23	0.2239	0.5408	104.46	113.47	0.2205
100	0.6579	109.30	119.04	0.2327	0.5751	108.82	118.39	0.2295
120	0.6927	113.56	123.82	0.2411	0.6073	113.15	123.27	0.2380
140	0.7261	117.85	128.60	0.2492	0.6380	117.50	128.12	0.2463
160	0.7584	122.18	133.41	0.2570	0.6675	121.87	132.98	0.2542
180	0.7898	126.55	138.25	0.2647	0.6961	126.28	137.87	0.2620
200	0.8205	130.98	143.13	0.2722	0.7239	130.73	142.79	0.2696
220	0.8506	135.47	148.06	0.2796	0.7512	135.25	147.76	0.2770
240	0.8803	140.02	153.05	0.2868	0.7779	139.82	152.77	0.2843
260	0.9095	144.63	158.10	0.2940	0.8043	144.45	157.84	0.2914
280	0.9384	149.32	163.21	0.3010	0.8303	149.15	162.97	0.2984
300	0.9671	154.06	168.38	0.3079	0.8561	153.91	168.16	0.3054
320	0.9955	158.88	173.62	0.3147	0.8816	158.73	173.42	0.3122
	$p = 100$ lbf/in.² ($T_{sat} = 79.17°F$)				$p = 120$ lbf/in.² ($T_{sat} = 90.54°F$)			
Sat.	0.4747	103.68	112.46	0.2169	0.3941	105.06	113.82	0.2165
80	0.4761	103.87	112.68	0.2173				
100	0.5086	108.32	117.73	0.2265	0.4080	107.26	116.32	0.2210
120	0.5388	112.73	122.70	0.2352	0.4355	111.84	121.52	0.2301
140	0.5674	117.13	127.63	0.2436	0.4610	116.37	126.61	0.2387
160	0.5947	121.55	132.55	0.2517	0.4852	120.89	131.66	0.2470
180	0.6210	125.99	137.49	0.2595	0.5082	125.42	136.70	0.2550
200	0.6466	130.48	142.45	0.2671	0.5305	129.97	141.75	0.2628
220	0.6716	135.02	147.45	0.2746	0.5520	134.56	146.82	0.2704
240	0.6960	139.61	152.49	0.2819	0.5731	139.20	151.92	0.2778
260	0.7201	144.26	157.59	0.2891	0.5937	143.89	157.07	0.2850
280	0.7438	148.98	162.74	0.2962	0.6140	148.63	162.26	0.2921
300	0.7672	153.75	167.95	0.3031	0.6339	153.43	167.51	0.2991
320	0.7904	158.59	173.21	0.3099	0.6537	158.29	172.81	0.3060

TABLE A-12E (*Continued*)

T °F	v ft³/lb	u Btu/lb	h Btu/lb	s Btu/lb · °R	v ft³/lb	u Btu/lb	h Btu/lb	s Btu/lb · °R
	$p = 140$ lbf/in.² ($T_{sat} = 100.56°F$)				$p = 160$ lbf/in.² ($T_{sat} = 109.55°F$)			
Sat.	0.3358	106.25	114.95	0.2161	0.2916	107.28	115.91	0.2157
120	0.3610	110.90	120.25	0.2254	0.3044	109.88	118.89	0.2209
140	0.3846	115.58	125.54	0.2344	0.3269	114.73	124.41	0.2303
160	0.4066	120.21	130.74	0.2429	0.3474	119.49	129.78	0.2391
180	0.4274	124.82	135.89	0.2511	0.3666	124.20	135.06	0.2475
200	0.4474	129.44	141.03	0.2590	0.3849	128.90	140.29	0.2555
220	0.4666	134.09	146.18	0.2667	0.4023	133.61	145.52	0.2633
240	0.4852	138.77	151.34	0.2742	0.4192	138.34	150.75	0.2709
260	0.5034	143.50	156.54	0.2815	0.4356	143.11	156.00	0.2783
280	0.5212	148.28	161.78	0.2887	0.4516	147.92	161.29	0.2856
300	0.5387	153.11	167.06	0.2957	0.4672	152.78	166.61	0.2927
320	0.5559	157.99	172.39	0.3026	0.4826	157.69	171.98	0.2996
340	0.5730	162.93	177.78	0.3094	0.4978	162.65	177.39	0.3065
360	0.5898	167.93	183.21	0.3162	0.5128	167.67	182.85	0.3132
	$p = 180$ lbf/in.² ($T_{sat} = 117.74°F$)				$p = 200$ lbf/in.² ($T_{sat} = 125.28°F$)			
Sat.	0.2569	108.18	116.74	0.2154	0.2288	108.98	117.44	0.2151
120	0.2595	108.77	117.41	0.2166				
140	0.2814	113.83	123.21	0.2264	0.2446	112.87	121.92	0.2226
160	0.3011	118.74	128.77	0.2355	0.2636	117.94	127.70	0.2321
180	0.3191	123.56	134.19	0.2441	0.2809	122.88	133.28	0.2410
200	0.3361	128.34	139.53	0.2524	0.2970	127.76	138.75	0.2494
220	0.3523	133.11	144.84	0.2603	0.3121	132.60	144.15	0.2575
240	0.3678	137.90	150.15	0.2680	0.3266	137.44	149.53	0.2653
260	0.3828	142.71	155.46	0.2755	0.3405	142.30	154.90	0.2728
280	0.3974	147.55	160.79	0.2828	0.3540	147.18	160.28	0.2802
300	0.4116	152.44	166.15	0.2899	0.3671	152.10	165.69	0.2874
320	0.4256	157.38	171.55	0.2969	0.3799	157.07	171.13	0.2945
340	0.4393	162.36	177.00	0.3038	0.3926	162.07	176.60	0.3014
360	0.4529	167.40	182.49	0.3106	0.4050	167.13	182.12	0.3082
	$p = 300$ lbf/in.² ($T_{sat} = 156.17°F$)				$p = 400$ lbf/in.² ($T_{sat} = 179.95°F$)			
Sat.	0.1424	111.72	119.62	0.2132	0.0965	112.77	119.91	0.2102
160	0.1462	112.95	121.07	0.2155				
180	0.1633	118.93	128.00	0.2265	0.0965	112.79	119.93	0.2102
200	0.1777	124.47	134.34	0.2363	0.1143	120.14	128.60	0.2235
220	0.1905	129.79	140.36	0.2453	0.1275	126.35	135.79	0.2343
240	0.2021	134.99	146.21	0.2537	0.1386	132.12	142.38	0.2438
260	0.2130	140.12	151.95	0.2618	0.1484	137.65	148.64	0.2527
280	0.2234	145.23	157.63	0.2696	0.1575	143.06	154.72	0.2610
300	0.2333	150.33	163.28	0.2772	0.1660	148.39	160.67	0.2689
320	0.2428	155.44	168.92	0.2845	0.1740	153.69	166.57	0.2766
340	0.2521	160.57	174.56	0.2916	0.1816	158.97	172.42	0.2840
360	0.2611	165.74	180.23	0.2986	0.1890	164.26	178.26	0.2912
380	0.2699	170.94	185.92	0.3055	0.1962	169.57	184.09	0.2983
400	0.2786	176.18	191.64	0.3122	0.2032	174.90	189.94	0.3051

TABLE A-13E Specific Heats of Selected Liquids and Solids

A. Liquid	State	c (Btu/lb $\cdot$ °R)
Ammonia	sat. liq., 0°F	1.08
	sat. liq., 50°F	1.12
	sat. liq., 120°F	1.22
Ethyl alcohol	1 atm, 76°F	0.580
Glycerin	1 atm, 50°F	0.554
	1 atm, 120°F	0.617
Mercury	1 atm, 50°F	0.033
	1 atm, 600°F	0.032
Refrigerant 12	sat. liq., 0°F	0.216
	sat. liq., 70°F	0.232
Water	1 atm, 32°F	1.007
	1 atm, 80°F	0.998
	1 atm, 212°F	1.007
B. Solid	**Temperature (°R)**	c (Btu/lb $\cdot$ °R)
Aluminum	540	0.216
Copper	540	0.0920
	720	0.0939
Ice	360	0.373
	460	0.471
	492	0.504
Iron	540	0.107
Lead	540	0.0308
Silver	540	0.0561

TABLE A-14E Ideal Gas Specific Heats of Some Common Gases (Btu/lb · °R)

Temp. °F	c_p	c_v	k	c_p	c_v	k	c_p	c_v	k	Temp. °F
	Air			Nitrogen, N_2			Oxygen, O_2			
40	0.240	0.171	1.401	0.248	0.177	1.400	0.219	0.156	1.397	40
100	0.240	0.172	1.400	0.248	0.178	1.399	0.220	0.158	1.394	100
200	0.241	0.173	1.397	0.249	0.178	1.398	0.223	0.161	1.387	200
300	0.243	0.174	1.394	0.250	0.179	1.396	0.226	0.164	1.378	300
400	0.245	0.176	1.389	0.251	0.180	1.393	0.230	0.168	1.368	400
500	0.248	0.179	1.383	0.254	0.183	1.388	0.235	0.173	1.360	500
600	0.250	0.182	1.377	0.256	0.185	1.383	0.239	0.177	1.352	600
700	0.254	0.185	1.371	0.260	0.189	1.377	0.242	0.181	1.344	700
800	0.257	0.188	1.365	0.262	0.191	1.371	0.246	0.184	1.337	800
900	0.259	0.191	1.358	0.265	0.194	1.364	0.249	0.187	1.331	900
1000	0.263	0.195	1.353	0.269	0.198	1.359	0.252	0.190	1.326	1000
1500	0.276	0.208	1.330	0.283	0.212	1.334	0.263	0.201	1.309	1500
2000	0.286	0.217	1.312	0.293	0.222	1.319	0.270	0.208	1.298	2000

Temp. °F	c_p	c_v	k	c_p	c_v	k	c_p	c_v	k	Temp. °F
	Carbon dioxide, CO_2			Carbon monoxide, CO			Hydrogen, H_2			
40	0.195	0.150	1.300	0.248	0.177	1.400	3.397	2.412	1.409	40
100	0.205	0.160	1.283	0.249	0.178	1.399	3.426	2.441	1.404	100
200	0.217	0.172	1.262	0.249	0.179	1.397	3.451	2.466	1.399	200
300	0.229	0.184	1.246	0.251	0.180	1.394	3.461	2.476	1.398	300
400	0.239	0.193	1.233	0.253	0.182	1.389	3.466	2.480	1.397	400
500	0.247	0.202	1.223	0.256	0.185	1.384	3.469	2.484	1.397	500
600	0.255	0.210	1.215	0.259	0.188	1.377	3.473	2.488	1.396	600
700	0.262	0.217	1.208	0.262	0.191	1.371	3.477	2.492	1.395	700
800	0.269	0.224	1.202	0.266	0.195	1.364	3.494	2.509	1.393	800
900	0.275	0.230	1.197	0.269	0.198	1.357	3.502	2.519	1.392	900
1000	0.280	0.235	1.192	0.273	0.202	1.351	3.513	2.528	1.390	1000
1500	0.298	0.253	1.178	0.287	0.216	1.328	3.618	2.633	1.374	1500
2000	0.312	0.267	1.169	0.297	0.226	1.314	3.758	2.773	1.355	2000

Source: Adapted from K. Wark, *Thermodynamics, 4th ed.*, McGraw-Hill, New York, 1983, as based on "Tables of Thermal Properties of Gases," NBS Circular 564, 1955.

TABLE A-15E Variation of $\bar{c}_p$ with Temperature for Various Ideal Gases

$$\frac{\bar{c}_p}{\bar{R}} = \alpha + \beta T + \gamma T^2 + \delta T^3 + \varepsilon T^4$$

T is in °R, equations valid from 540 to 1800 °R

Gas	α	$\beta \times 10^3$	$\gamma \times 10^6$	$\delta \times 10^9$	$\epsilon \times 10^{12}$
CO	3.710	−0.899	1.140	−0.348	0.0228
CO_2	2.401	4.853	−2.039	0.343	0
H_2	3.057	1.487	−1.793	0.947	−0.1726
H_2O	4.070	−0.616	1.281	−0.508	0.0769
O_2	3.626	−1.043	2.178	−1.160	0.2053
N_2	3.675	−0.671	0.717	−0.108	−0.0215
Air	3.653	−0.7428	1.017	−0.328	0.02632
NH_3	3.591	0.274	2.576	−1.437	0.2601
NO	4.046	−1.899	2.464	−1.048	0.1517
NO_2	3.459	1.147	2.064	−1.639	0.3448
SO_2	3.267	2.958	0.211	−0.906	0.2438
SO_3	2.578	8.087	−2.832	−0.136	0.1878
CH_4	3.826	−2.211	7.580	−3.898	0.6633
C_2H_2	1.410	10.587	−7.562	2.811	−0.3939
C_2H_4	1.426	6.324	2.466	−2.787	0.6429
Monatomic gases[a]	2.5	0	0	0	0

[a] For monatomic gases, such as He, Ne, and Ar, $\bar{c}_p$ is constant over a wide temperature range and is very nearly equal to $5/2 \bar{R}$.

Source: Adapted from K. Wark, *Thermodynamics, 4th ed.,* McGraw-Hill, New York, 1983, as based on NASA SP-273, U.S. Government Printing Office, Washington, DC, 1971.

TABLE A-16E Ideal Gas Properties of Air

$T(°R)$, h and u(Btu/lb), $s°$(Btu/lb·°R)

T	h	p_r	u	v_r	$s°$	T	h	p_r	u	v_r	$s°$
360	85.97	0.3363	61.29	396.6	0.50369	940	226.11	9.834	161.68	35.41	0.73509
380	90.75	0.4061	64.70	346.6	0.51663	960	231.06	10.61	165.26	33.52	0.74030
400	95.53	0.4858	68.11	305.0	0.52890	980	236.02	11.43	168.83	31.76	0.74540
420	100.32	0.5760	71.52	270.1	0.54058	1000	240.98	12.30	172.43	30.12	0.75042
440	105.11	0.6776	74.93	240.6	0.55172	1040	250.95	14.18	179.66	27.17	0.76019
460	109.90	0.7913	78.36	215.33	0.56235	1080	260.97	16.28	186.93	24.58	0.76964
480	114.69	0.9182	81.77	193.65	0.57255	1120	271.03	18.60	194.25	22.30	0.77880
500	119.48	1.0590	85.20	174.90	0.58233	1160	281.14	21.18	201.63	20.29	0.78767
520	124.27	1.2147	88.62	158.58	0.59172	1200	291.30	24.01	209.05	18.51	0.79628
537	128.34	1.3593	91.53	146.34	0.59945	1240	301.52	27.13	216.53	16.93	0.80466
540	129.06	1.3860	92.04	144.32	0.60078	1280	311.79	30.55	224.05	15.52	0.81280
560	133.86	1.5742	95.47	131.78	0.60950	1320	322.11	34.31	231.63	14.25	0.82075
580	138.66	1.7800	98.90	120.70	0.61793	1360	332.48	38.41	239.25	13.12	0.82848
600	143.47	2.005	102.34	110.88	0.62607	1400	342.90	42.88	246.93	12.10	0.83604
620	148.28	2.249	105.78	102.12	0.63395	1440	353.37	47.75	254.66	11.17	0.84341
640	153.09	2.514	109.21	94.30	0.64159	1480	363.89	53.04	262.44	10.34	0.85062
660	157.92	2.801	112.67	87.27	0.64902	1520	374.47	58.78	270.26	9.578	0.85767
680	162.73	3.111	116.12	80.96	0.65621	1560	385.08	65.00	278.13	8.890	0.86456
700	167.56	3.446	119.58	75.25	0.66321	1600	395.74	71.73	286.06	8.263	0.87130
720	172.39	3.806	123.04	70.07	0.67002	1650	409.13	80.89	296.03	7.556	0.87954
740	177.23	4.193	126.51	65.38	0.67665	1700	422.59	90.95	306.06	6.924	0.88758
760	182.08	4.607	129.99	61.10	0.68312	1750	436.12	101.98	316.16	6.357	0.89542
780	186.94	5.051	133.47	57.20	0.68942	1800	449.71	114.0	326.32	5.847	0.90308
800	191.81	5.526	136.97	53.63	0.69558	1850	463.37	127.2	336.55	5.388	0.91056
820	196.69	6.033	140.47	50.35	0.70160	1900	477.09	141.5	346.85	4.974	0.91788
840	201.56	6.573	143.98	47.34	0.70747	1950	490.88	157.1	357.20	4.598	0.92504
860	206.46	7.149	147.50	44.57	0.71323	2000	504.71	174.0	367.61	4.258	0.93205
880	211.35	7.761	151.02	42.01	0.71886	2050	518.61	192.3	378.08	3.949	0.93891
900	216.26	8.411	154.57	39.64	0.72438	2100	532.55	212.1	388.60	3.667	0.94564
920	221.18	9.102	158.12	37.44	0.72979	2150	546.54	233.5	399.17	3.410	0.95222

TABLE A-16E *(Continued)*

$T(°R)$, h and u(Btu/lb), s^o(Btu/lb · °R)

T	h	p_r	u	v_r	s^o		T	h	p_r	u	v_r	s^o
2200	560.59	256.6	409.78	3.176	0.95868		3700	998.11	2330	744.48	.5882	1.10991
2250	574.69	281.4	420.46	2.961	0.96501		3750	1013.1	2471	756.04	.5621	1.11393
2300	588.82	308.1	431.16	2.765	0.97123		3800	1028.1	2618	767.60	.5376	1.11791
2350	603.00	336.8	441.91	2.585	0.97732		3850	1043.1	2773	779.19	.5143	1.12183
2400	617.22	367.6	452.70	2.419	0.98331		3900	1058.1	2934	790.80	.4923	1.12571
2450	631.48	400.5	463.54	2.266	0.98919		3950	1073.2	3103	802.43	.4715	1.12955
2500	645.78	435.7	474.40	2.125	0.99497		4000	1088.3	3280	814.06	.4518	1.13334
2550	660.12	473.3	485.31	1.996	1.00064		4050	1103.4	3464	825.72	.4331	1.13709
2600	674.49	513.5	496.26	1.876	1.00623		4100	1118.5	3656	837.40	.4154	1.14079
2650	688.90	556.3	507.25	1.765	1.01172		4150	1133.6	3858	849.09	.3985	1.14446
2700	703.35	601.9	518.26	1.662	1.01712		4200	1148.7	4067	860.81	.3826	1.14809
2750	717.83	650.4	529.31	1.566	1.02244		4300	1179.0	4513	884.28	.3529	1.15522
2800	732.33	702.0	540.40	1.478	1.02767		4400	1209.4	4997	907.81	.3262	1.16221
2850	746.88	756.7	551.52	1.395	1.03282		4500	1239.9	5521	931.39	.3019	1.16905
2900	761.45	814.8	562.66	1.318	1.03788		4600	1270.4	6089	955.04	.2799	1.17575
2950	776.05	876.4	573.84	1.247	1.04288		4700	1300.9	6701	978.73	.2598	1.18232
3000	790.68	941.4	585.04	1.180	1.04779		4800	1331.5	7362	1002.5	.2415	1.18876
3050	805.34	1011	596.28	1.118	1.05264		4900	1362.2	8073	1026.3	.2248	1.19508
3100	820.03	1083	607.53	1.060	1.05741		5000	1392.9	8837	1050.1	.2096	1.20129
3150	834.75	1161	618.82	1.006	1.06212		5100	1423.6	9658	1074.0	.1956	1.20738
3200	849.48	1242	630.12	.9546	1.06676		5200	1454.4	10539	1098.0	.1828	1.21336
3250	864.24	1328	641.46	.9069	1.07134		5300	1485.3	11481	1122.0	.1710	1.21923
3300	879.02	1418	652.81	.8621	1.07585							
3350	893.83	1513	664.20	.8202	1.08031							
3400	908.66	1613	675.60	.7807	1.08470							
3450	923.52	1719	687.04	.7436	1.08904							
3500	938.40	1829	698.48	.7087	1.09332							
3550	953.30	1946	709.95	.6759	1.09755							
3600	968.21	2068	721.44	.6449	1.10172							
3650	983.15	2196	732.95	.6157	1.10584							

Source: Tables A-13E through A-19E are adapted from K. Wark, *Thermodynamics, 4th ed.*, McGraw-Hill, New York, 1983, as extracted from J. H. Keenan and J. Kaye, "Gas Tables," Wiley, New York, 1945.

TABLE A-17E Ideal Gas Properties of Nitrogen, N_2

$T(°R)$, $\bar{h}$ and $\bar{u}$(Btu/lbmol), $\bar{s}°$(Btu/lbmol · °R)

$[\bar{h}_f° = 0$ Btu/lbmol]

T	$\bar{h}$	$\bar{u}$	$\bar{s}°$	T	$\bar{h}$	$\bar{u}$	$\bar{s}°$
300	2082.0	1486.2	41.695	1080	7551.0	5406.2	50.651
320	2221.0	1585.5	42.143	1100	7695.0	5510.5	50.783
340	2360.0	1684.4	42.564	1120	7839.3	5615.2	50.912
360	2498.9	1784.0	42.962	1140	7984.0	5720.1	51.040
380	2638.0	1883.4	43.337	1160	8129.0	5825.4	51.167
400	2777.0	1982.6	43.694	1180	8274.4	5931.0	51.291
420	2916.1	2082.0	44.034	1200	8420.0	6037.0	51.413
440	3055.1	2181.3	44.357	1220	8566.1	6143.4	51.534
460	3194.1	2280.6	44.665	1240	8712.6	6250.1	51.653
480	3333.1	2379.9	44.962	1260	8859.3	6357.2	51.771
500	3472.2	2479.3	45.246	1280	9006.4	6464.5	51.887
520	3611.3	2578.6	45.519	1300	9153.9	6572.3	51.001
537	3729.5	2663.1	45.743	1320	9301.8	6680.4	52.114
540	3750.3	2678.0	45.781	1340	9450.0	6788.9	52.225
560	3889.5	2777.4	46.034	1360	9598.6	6897.8	52.335
580	4028.7	2876.9	46.278	1380	9747.5	7007.0	52.444
600	4167.9	2976.4	46.514	1400	9896.9	7116.7	52.551
620	4307.1	3075.9	46.742	1420	10046.6	7226.7	52.658
640	4446.4	3175.5	46.964	1440	10196.6	7337.0	52.763
660	4585.8	3275.2	47.178	1460	10347.0	7447.6	52.867
680	4725.3	3374.9	47.386	1480	10497.8	7558.7	52.969
700	4864.9	3474.8	47.588	1500	10648.0	7670.1	53.071
720	5004.5	3574.7	47.785	1520	10800.4	7781.9	53.171
740	5144.3	3674.7	47.977	1540	10952.2	7893.9	53.271
760	5284.1	3774.9	48.164	1560	11104.3	8006.4	53.369
780	5424.2	3875.2	48.345	1580	11256.9	8119.2	53.465
800	5564.4	3975.7	48.522	1600	11409.7	8232.3	53.561
820	5704.7	4076.3	48.696	1620	11562.8	8345.7	53.656
840	5845.3	4177.1	48.865	1640	11716.4	8459.6	53.751
860	5985.9	4278.1	49.031	1660	11870.2	8573.6	53.844
880	6126.9	4379.4	49.193	1680	12024.3	8688.1	53.936
900	6268.1	4480.8	49.352	1700	12178.9	8802.9	54.028
920	6409.6	4582.6	49.507	1720	12333.7	8918.0	54.118
940	6551.2	4684.5	49.659	1740	12488.8	9033.4	54.208
960	6693.1	4786.7	49.808	1760	12644.3	9149.2	54.297
980	6835.4	4889.3	49.955	1780	12800.2	9265.3	54.385
1000	6977.9	4992.0	50.099	1800	12956.3	9381.7	54.472
1020	7120.7	5095.1	50.241	1820	13112.7	9498.4	54.559
1040	7263.8	5198.5	50.380	1840	13269.5	9615.5	54.645
1060	7407.2	5302.2	50.516	1860	13426.5	9732.8	54.729

TABLE A-17E (*Continued*)

T	$\bar{h}$	$\bar{u}$	$\bar{s}°$	T	$\bar{h}$	$\bar{u}$	$\bar{s}°$
1900	13,742	9,968	54.896	3500	27,016	20,065	59.944
1940	14,058	10,205	55.061	3540	27,359	20,329	60.041
1980	14,375	10,443	55.223	3580	27,703	20,593	60.138
2020	14,694	10,682	55.383	3620	28,046	20,858	60.234
2060	15,013	10,923	55.540	3660	28,391	21,122	60.328
2100	15,334	11,164	55.694	3700	28,735	21,387	60.422
2140	15,656	11,406	55.846	3740	29,080	21,653	60.515
2180	15,978	11,649	55.995	3780	29,425	21,919	60.607
2220	16,302	11,893	56.141	3820	29,771	22,185	60.698
2260	16,626	12,138	56.286	3860	30,117	22,451	60.788
2300	16,951	12,384	56.429	3900	30,463	22,718	60.877
2340	17,277	12,630	56.570	3940	30,809	22,985	60.966
2380	17,604	12,878	56.708	3980	31,156	23,252	61.053
2420	17,932	13,126	56.845	4020	31,503	23,520	61.139
2460	18,260	13,375	56.980	4060	31,850	23,788	61.225
2500	18,590	13,625	57.112	4100	32,198	24,056	61.310
2540	18,919	13,875	57.243	4140	32,546	24,324	61.395
2580	19,250	14,127	57.372	4180	32,894	24,593	61.479
2620	19,582	14,379	57.499	4220	33,242	24,862	61.562
2660	19,914	14,631	57.625	4260	33,591	25,131	61.644
2700	20,246	14,885	57.750	4300	33,940	25,401	61.726
2740	20,580	15,139	57.872	4340	34,289	25,670	61.806
2780	20,914	15,393	57.993	4380	34,638	25,940	61.887
2820	21,248	15,648	58.113	4420	34,988	26,210	61.966
2860	21,584	15,905	58.231	4460	35,338	26,481	62.045
2900	21,920	16,161	58.348	4500	35,688	26,751	62.123
2940	22,256	16,417	58.463	4540	36,038	27,022	62.201
2980	22,593	16,675	58.576	4580	36,389	27,293	62.278
3020	22,930	16,933	58.688	4620	36,739	27,565	62.354
3060	23,268	17,192	58.800	4660	37,090	27,836	62.429
3100	23,607	17,451	58.910	4700	37,441	28,108	62.504
3140	23,946	17,710	59.019	4740	37,792	28,379	62.578
3180	24,285	17,970	59.126	4780	38,144	28,651	62.652
3220	24,625	18,231	59.232	4820	38,495	28,924	62.725
3260	24,965	18,491	59.338	4860	38,847	29,196	62.798
3300	25,306	18,753	59.442	4900	39,199	29,468	62.870
3340	25,647	19,014	59.544	5000	40,080	30,151	63.049
3380	25,989	19,277	59.646	5100	40,962	30,834	63.223
3420	26,331	19,539	59.747	5200	41,844	31,518	63.395
3460	26,673	19,802	59.846	5300	42,728	32,203	63.563

TABLE A-18E Ideal Gas Properties of Oxygen, O_2

$T(°R)$, $\bar{h}$ and $\bar{u}(Btu/lbmol)$, $\bar{s}°(Btu/lbmol \cdot °R)$

$[\bar{h}_f° = 0\ Btu/lbmol]$

T	$\bar{h}$	$\bar{u}$	$\bar{s}°$	T	$\bar{h}$	$\bar{u}$	$\bar{s}°$
300	2073.5	1477.8	44.927	1080	7696.8	5552.1	54.064
320	2212.6	1577.1	45.375	1100	7850.4	5665.9	54.204
340	2351.7	1676.5	45.797	1120	8004.5	5780.3	54.343
360	2490.8	1775.9	46.195	1140	8159.1	5895.2	54.480
380	2630.0	1875.3	46.571	1160	8314.2	6010.6	54.614
400	2769.1	1974.8	46.927	1180	8469.8	6126.5	54.748
420	2908.3	2074.3	47.267	1200	8625.8	6242.8	54.879
440	3047.5	2173.8	47.591	1220	8782.4	6359.6	55.008
460	3186.9	2273.4	47.900	1240	8939.4	6476.9	55.136
480	3326.5	2373.3	48.198	1260	9096.7	6594.5	55.262
500	3466.2	2473.2	48.483	1280	9254.6	6712.7	55.386
520	3606.1	2573.4	48.757	1300	9412.9	6831.3	55.508
537	3725.1	2658.7	48.982	1320	9571.6	6950.2	55.630
540	3746.2	2673.8	49.021	1340	9730.7	7069.6	55.750
560	3886.6	2774.5	49.276	1360	9890.2	7189.4	55.867
580	4027.3	2875.5	49.522	1380	10050.1	7309.6	55.984
600	4168.3	2976.8	49.762	1400	10210.4	7430.1	56.099
620	4309.7	3078.4	49.993	1420	10371.0	7551.1	56.213
640	4451.4	3180.4	50.218	1440	10532.0	7672.4	56.326
660	4593.5	3282.9	50.437	1460	10693.3	7793.9	56.437
680	4736.2	3385.8	50.650	1480	10855.1	7916.0	56.547
700	4879.3	3489.2	50.858	1500	11017.1	8038.3	56.656
720	5022.9	3593.1	51.059	1520	11179.6	8161.1	56.763
740	5167.0	3697.4	51.257	1540	11342.4	8284.2	56.869
760	5311.4	3802.2	51.450	1560	11505.4	8407.4	56.975
780	5456.4	3907.5	51.638	1580	11668.8	8531.1	57.079
800	5602.0	4013.3	51.821	1600	11832.5	8655.1	57.182
820	5748.1	4119.7	52.002	1620	11996.6	8779.5	57.284
840	5894.8	4226.6	52.179	1640	12160.9	8904.1	57.385
860	6041.9	4334.1	52.352	1660	12325.5	9029.0	57.484
880	6189.6	4442.0	52.522	1680	12490.4	9154.1	57.582
900	6337.9	4550.6	52.688	1700	12655.6	9279.6	57.680
920	6486.7	4659.7	52.852	1720	12821.1	9405.4	57.777
940	6636.1	4769.4	53.012	1740	12986.9	9531.5	57.873
960	6786.0	4879.5	53.170	1760	13153.0	9657.9	57.968
980	6936.4	4990.3	53.326	1780	13319.2	9784.4	58.062
1000	7087.5	5101.6	53.477	1800	13485.8	9911.2	58.155
1020	7238.9	5213.3	53.628	1820	13652.5	10038.2	58.247
1040	7391.0	5325.7	53.775	1840	13819.6	10165.6	58.339
1060	7543.6	5438.6	53.921	1860	13986.8	10293.1	58.428

TABLE A-18E (*Continued*)

T	$\bar{h}$	$\bar{u}$	$\bar{s}^{\circ}$	T	$\bar{h}$	$\bar{u}$	$\bar{s}^{\circ}$
1900	14,322	10,549	58.607	3500	28,273	21,323	63.914
1940	14,658	10,806	58.782	3540	28,633	21,603	64.016
1980	14,995	11,063	58.954	3580	28,994	21,884	64.114
2020	15,333	11,321	59.123	3620	29,354	22,165	64.217
2060	15,672	11,581	59.289	3660	29,716	22,447	64.316
2100	16,011	11,841	59.451	3700	30,078	22,730	64.415
2140	16,351	12,101	59.612	3740	30,440	23,013	64.512
2180	16,692	12,363	59.770	3780	30,803	23,296	64.609
2220	17,036	12,625	59.926	3820	31,166	23,580	64.704
2260	17,376	12,888	60.077	3860	31,529	23,864	64.800
2300	17,719	13,151	60.228	3900	31,894	24,149	64.893
2340	18,062	13,416	60.376	3940	32,258	24,434	64.986
2380	18,407	13,680	60.522	3980	32,623	24,720	65.078
2420	18,572	13,946	60.666	4020	32,989	25,006	65.169
2460	19,097	14,212	60.808	4060	33,355	25,292	65.260
2500	19,443	14,479	60.946	4100	33,722	25,580	65.350
2540	19,790	14,746	61.084	4140	34,089	25,867	64.439
2580	20,138	15,014	61.220	4180	34,456	26,155	65.527
2620	20,485	15,282	61.354	4220	34,824	26,444	65.615
2660	20,834	15,551	61.486	4260	35,192	26,733	65.702
2700	21,183	15,821	61.616	4300	35,561	27,022	65.788
2740	21,533	16,091	61.744	4340	35,930	27,312	65.873
2780	21,883	16,362	61.871	4380	36,300	27,602	65.958
2820	22,232	16,633	61.996	4420	36,670	27,823	66.042
2860	22,584	16,905	62.120	4460	37,041	28,184	66.125
2900	22,936	17,177	62.242	4500	37,412	28,475	66.208
2940	23,288	17,450	62.363	4540	37,783	28,768	66.290
2980	23,641	17,723	62.483	4580	38,155	29,060	66.372
3020	23,994	17,997	62.599	4620	38,528	29,353	66.453
3060	24,348	18,271	62.716	4660	38,900	29,646	66.533
3100	24,703	18,546	62.831	4700	39,274	29,940	66.613
3140	25,057	18,822	62.945	4740	39,647	30,234	66.691
3180	25,413	19,098	63.057	4780	40,021	30,529	66.770
3220	25,769	19,374	63.169	4820	40,396	30,824	66.848
3260	26,175	19,651	63.279	4860	40,771	31,120	66.925
3300	26,412	19,928	63.386	4900	41,146	31,415	67.003
3340	26,839	20,206	63.494	5000	42,086	32,157	67.193
3380	27,197	20,485	63.601	5100	43,021	32,901	67.380
3420	27,555	20,763	63.706	5200	43,974	33,648	67.562
3460	27,914	21,043	63.811	5300	44,922	34,397	67.743

TABLE A-19E Ideal Gas Properties of Water Vapor, H_2O

$T(°R)$, $\bar{h}$ and $\bar{u}$(Btu/lbmol), $\bar{s}°$(Btu/lbmol · °R)

$[\bar{h}_f° = -104{,}040 \text{ Btu/lbmol}]$

T	$\bar{h}$	$\bar{u}$	$\bar{s}°$	T	$\bar{h}$	$\bar{u}$	$\bar{s}°$
300	2367.6	1771.8	40.439	1080	8768.2	6623.5	50.854
320	2526.8	1891.3	40.952	1100	8942.0	6757.5	51.013
340	2686.0	2010.8	41.435	1120	9116.4	6892.2	51.171
360	2845.1	2130.2	41.889	1140	9291.4	7027.5	51.325
380	3004.4	2249.8	42.320	1160	9467.1	7163.5	51.478
400	3163.8	2369.4	42.728	1180	9643.4	7300.1	51.630
420	3323.2	2489.1	43.117	1200	9820.4	7437.4	51.777
440	3482.7	2608.9	43.487	1220	9998.0	7575.2	51.925
460	3642.3	2728.8	43.841	1240	10176.1	7713.6	52.070
480	3802.0	2848.8	44.182	1260	10354.9	7852.7	52.212
500	3962.0	2969.1	44.508	1280	10534.4	7992.5	52.354
520	4122.0	3089.4	44.821	1300	10714.5	8132.9	52.494
537	4258.0	3191.9	45.079	1320	10895.3	8274.0	52.631
540	4282.4	3210.0	45.124	1340	11076.6	8415.5	52.768
560	4442.8	3330.7	45.415	1360	11258.7	8557.9	52.903
580	4603.7	3451.9	45.696	1380	11441.4	8700.9	53.037
600	4764.7	3573.2	45.970	1400	11624.8	8844.6	53.168
620	4926.1	3694.9	46.235	1420	11808.8	8988.9	53.299
640	5087.8	3816.8	46.492	1440	11993.4	9133.8	53.428
660	5250.0	3939.3	46.741	1460	12178.8	9279.4	53.556
680	5412.5	4062.1	46.984	1480	12364.8	9425.7	53.682
700	5575.4	4185.3	47.219	1500	12551.4	9572.7	53.808
720	5738.8	4309.0	47.450	1520	12738.8	9720.3	53.932
740	5902.6	4433.1	47.673	1540	12926.8	9868.6	54.055
760	6066.9	4557.6	47.893	1560	13115.6	10017.6	54.117
780	6231.7	4682.7	48.106	1580	13305.0	10167.3	54.298
800	6396.9	4808.2	48.316	1600	13494.4	10317.6	54.418
820	6562.6	4934.2	48.520	1620	13685.7	10468.6	54.535
840	6728.9	5060.8	48.721	1640	13877.0	10620.2	54.653
860	6895.6	5187.8	48.916	1660	14069.2	10772.7	54.770
880	7062.9	5315.3	49.109	1680	14261.9	10925.6	54.886
900	7230.9	5443.6	49.298	1700	14455.4	11079.4	54.999
920	7399.4	5572.4	49.483	1720	14649.5	11233.8	55.113
940	7568.4	5701.7	49.665	1740	14844.3	11388.9	55.226
960	7738.0	5831.6	49.843	1760	15039.8	11544.7	55.339
980	7908.2	5962.0	50.019	1780	15236.1	11701.2	55.449
1000	8078.9	6093.0	50.191	1800	15433.0	11858.4	55.559
1020	8250.4	6224.8	50.360	1820	15630.6	12016.3	55.668
1040	8422.4	6357.1	50.528	1840	15828.7	12174.7	55.777
1060	8595.0	6490.0	50.693	1860	16027.6	12333.9	55.884

TABLE A-19E (*Continued*)

T	$\bar{h}$	$\bar{u}$	$\bar{s}°$	T	$\bar{h}$	$\bar{u}$	$\bar{s}°$
1900	16,428	12,654	56.097	3500	34,324	27,373	62.876
1940	16,830	12,977	56.307	3540	34,809	27,779	63.015
1980	17,235	13,303	56.514	3580	35,296	28,187	63.153
2020	17,643	13,632	56.719	3620	35,785	28,596	63.288
2060	18,054	13,963	56.920	3660	36,274	29,006	63.423
2100	18,467	14,297	57.119	3700	36,765	29,418	63.557
2140	18,883	14,633	57.315	3740	37,258	29,831	63.690
2180	19,301	14,972	57.509	3780	37,752	30,245	63.821
2220	19,722	15,313	57.701	3820	38,247	30,661	63.952
2260	20,145	15,657	57.889	3860	38,743	31,077	64.082
2300	20,571	16,003	58.077	3900	39,240	31,495	64.210
2340	20,999	16,352	58.261	3940	39,739	31,915	64.338
2380	21,429	16,703	58.445	3980	40,239	32,335	64.465
2420	21,862	17,057	58.625	4020	40,740	32,757	64.591
2460	22,298	17,413	58.803	4060	41,242	33,179	64.715
2500	22,735	17,771	58.980	4100	41,745	33,603	64.839
2540	23,175	18,131	59.155	4140	42,250	34,028	64.962
2580	23,618	18,494	59.328	4180	42,755	34,454	65.084
2620	24,062	18,859	59.500	4220	43,267	34,881	65.204
2660	24,508	19,226	59.669	4260	43,769	35,310	65.325
2700	24,957	19,595	59.837	4300	44,278	35,739	65.444
2740	25,408	19,967	60.003	4340	44,788	36,169	65.563
2780	25,861	20,340	60.167	4380	45,298	36,600	65.680
2820	26,316	20,715	60.330	4420	45,810	37,032	65.797
2860	26,773	21,093	60.490	4460	46,322	37,465	65.913
2900	27,231	21,472	60.650	4500	46,836	37,900	66.028
2940	27,692	21,853	60.809	4540	47,350	38,334	66.142
2980	28,154	22,237	60.965	4580	47,866	38,770	66.255
3020	28,619	22,621	61.120	4620	48,382	39,207	66.368
3060	29,085	23,085	61.274	4660	48,899	39,645	66.480
3100	29,553	23,397	61.426	4700	49,417	40,083	66.591
3140	30,023	23,787	61.577	4740	49,936	40,523	66.701
3180	30,494	24,179	61.727	4780	50,455	40,963	66.811
3220	30,967	24,572	61.874	4820	50,976	41,404	66.920
3260	31,442	24,968	62.022	4860	51,497	41,856	67.028
3300	31,918	25,365	62.167	4900	52,019	42,288	67.135
3340	32,396	25,763	62.312	5000	53,327	43,398	67.401
3380	32,876	26,164	62.454	5100	54,640	44,512	67.662
3420	33,357	26,565	62.597	5200	55,957	45,631	67.918
3460	33,839	26,968	62.738	5300	57,279	46,754	68.172

TABLE A-20E Ideal Gas Properties of Carbon Monoxide, CO

$T(°R)$, $\bar{h}$ and $\bar{u}$(Btu/lbmol), $\bar{s}°$(Btu/lbmol · °R)

$[\bar{h}_f° = -47{,}540 \text{ Btu/lbmol}]$

T	$\bar{h}$	$\bar{u}$	$\bar{s}°$	T	$\bar{h}$	$\bar{u}$	$\bar{s}°$
300	2081.9	1486.1	43.223	1080	7571.1	5426.4	52.203
320	2220.9	1585.4	43.672	1100	7716.8	5532.3	52.337
340	2359.9	1684.7	44.093	1120	7862.9	5638.7	52.468
360	2498.8	1783.9	44.490	1140	8009.2	5745.4	52.598
380	2637.9	1883.3	44.866	1160	8156.1	5851.5	52.726
400	2776.9	1982.6	45.223	1180	8303.3	5960.0	52.852
420	2916.0	2081.9	45.563	1200	8450.8	6067.8	52.976
440	3055.0	2181.2	45.886	1220	8598.8	6176.0	53.098
460	3194.0	2280.5	46.194	1240	8747.2	6284.7	53.218
480	3333.0	2379.8	46.491	1260	8896.0	6393.8	53.337
500	3472.1	2479.2	46.775	1280	9045.0	6503.1	53.455
520	3611.2	2578.6	47.048	1300	9194.6	6613.0	53.571
537	3725.1	2663.1	47.272	1320	9344.6	6723.2	53.685
540	3750.3	2677.9	47.310	1340	9494.8	6833.7	53.799
560	3889.5	2777.4	47.563	1360	9645.5	6944.7	53.910
580	4028.7	2876.9	47.807	1380	9796.6	7056.1	54.021
600	4168.0	2976.5	48.044	1400	9948.1	7167.9	54.129
620	4307.4	3076.2	48.272	1420	10100.0	7280.1	54.237
640	4446.9	3175.9	48.494	1440	10252.2	7392.6	54.344
660	4586.6	3275.8	48.709	1460	10404.8	7505.4	54.448
680	4726.2	3375.8	48.917	1480	10557.8	7618.7	54.522
700	4866.0	3475.9	49.120	1500	10711.1	7732.3	54.665
720	5006.1	3576.3	49.317	1520	10864.9	7846.4	54.757
740	5146.4	3676.9	49.509	1540	11019.0	7960.8	54.858
760	5286.8	3777.5	49.697	1560	11173.4	8075.4	54.958
780	5427.4	3878.4	49.880	1580	11328.2	8190.5	55.056
800	5568.2	3979.5	50.058	1600	11483.4	8306.0	55.154
820	5709.4	4081.0	50.232	1620	11638.9	8421.8	55.251
840	5850.7	4182.6	50.402	1640	11794.7	8537.9	55.347
860	5992.3	4284.5	50.569	1660	11950.9	8654.4	55.411
880	6134.2	4386.6	50.732	1680	12107.5	8771.2	55.535
900	6276.4	4489.1	50.892	1700	12264.3	8888.3	55.628
920	6419.0	4592.0	51.048	1720	12421.4	9005.7	55.720
940	6561.7	4695.0	51.202	1740	12579.0	9123.6	55.811
960	6704.9	4798.5	51.353	1760	12736.7	9241.6	55.900
980	6848.4	4902.3	51.501	1780	12894.9	9360.0	55.990
1000	6992.2	5006.3	51.646	1800	13053.2	9478.6	56.078
1020	7136.4	5110.8	51.788	1820	13212.0	9597.7	56.166
1040	7281.0	5215.7	51.929	1840	13371.0	9717.0	56.253
1060	7425.9	5320.9	52.067	1860	13530.2	9836.5	56.339

TABLE A-20E (*Continued*)

T	$\bar{h}$	$\bar{u}$	$\bar{s}\,°$	T	$\bar{h}$	$\bar{u}$	$\bar{s}\,°$
1900	13,850	10,077	56.509	3500	27,262	20,311	61.612
1940	14,170	10,318	56.677	3540	27,608	20,576	61.710
1980	14,492	10,560	56.841	3580	27,954	20,844	61.807
2020	14,815	10,803	57.007	3620	28,300	21,111	61.903
2060	15,139	11,048	57.161	3660	28,647	21,378	61.998
2100	15,463	11,293	57.317	3700	28,994	21,646	62.093
2140	15,789	11,539	57.470	3740	29,341	21,914	62.186
2180	16,116	11,787	57.621	3780	29,688	22,182	62.279
2220	16,443	12,035	57.770	3820	30,036	22,450	62.370
2260	16,722	12,284	57.917	3860	30,384	22,719	62.461
2300	17,101	12,534	58.062	3900	30,733	22,988	62.511
2340	17,431	12,784	58.204	3940	31,082	23,257	62.640
2380	17,762	13,035	58.344	3980	31,431	23,527	62.728
2420	18,093	13,287	58.482	4020	31,780	23,797	62.816
2460	18,426	13,541	58.619	4060	32,129	24,067	62.902
2500	18,759	13,794	58.754	4100	32,479	24,337	62.988
2540	19,093	14,048	58.885	4140	32,829	24,608	63.072
2580	19,427	14,303	59.016	4180	33,179	24,878	63.156
2620	19,762	14,559	59.145	4220	33,530	25,149	63.240
2660	20,098	14,815	59.272	4260	33,880	25,421	63.323
2700	20,434	15,072	59.398	4300	34,231	25,692	63.405
2740	20,771	15,330	59.521	4340	34,582	25,934	63.486
2780	21,108	15,588	59.644	4380	34,934	26,235	63.567
2820	21,446	15,846	59.765	4420	35,285	26,508	63.647
2860	21,785	16,105	59.884	4460	35,637	26,780	63.726
2900	22,124	16,365	60.002	4500	35,989	27,052	63.805
2940	22,463	16,625	60.118	4540	36,341	27,325	63.883
2980	22,803	16,885	60.232	4580	36,693	27,598	63.960
3020	23,144	17,146	60.346	4620	37,046	27,871	64.036
3060	23,485	17,408	60.458	4660	37,398	28,144	64.113
3100	23,826	17,670	60.569	4700	37,751	28,417	64.188
3140	24,168	17,932	60.679	4740	38,104	28,691	64.263
3180	24,510	18,195	60.787	4780	38,457	28,965	64.337
3220	24,853	18,458	60.894	4820	38,811	29,239	64.411
3260	25,196	18,722	61.000	4860	39,164	29,513	64.484
3300	25,539	18,986	61.105	4900	39,518	29,787	64.556
3340	25,883	19,250	61.209	5000	40,403	30,473	64.735
3380	26,227	19,515	61.311	5100	41,289	31,161	64.910
3420	26,572	19,780	61.412	5200	42,176	31,849	65.082
3460	26,917	20,045	61.513	5300	43,063	32,538	65.252

TABLE A-21E Ideal Gas Properties of Carbon Dioxide, CO_2

$T(°R)$, $\bar{h}$ and $\bar{u}$(Btu/lbmol), $\bar{s}°$(Btu/lbmol · °R)

$[\bar{h}_f° = -169,300 \text{ Btu/lbmol}]$

T	$\bar{h}$	$\bar{u}$	$\bar{s}°$	T	$\bar{h}$	$\bar{u}$	$\bar{s}°$
300	2108.2	1512.4	46.353	1080	9575.8	7431.1	58.072
320	2256.6	1621.1	46.832	1100	9802.6	7618.1	58.281
340	2407.3	1732.1	47.289	1120	10030.6	7806.4	58.485
360	2560.5	1845.6	47.728	1140	10260.1	7996.2	58.689
380	2716.4	1961.8	48.148	1160	10490.6	8187.0	58.889
400	2874.7	2080.4	48.555	1180	10722.3	8379.0	59.088
420	3035.7	2201.7	48.947	1200	10955.3	8572.3	59.283
440	3199.4	2325.6	49.329	1220	11189.4	8766.6	59.477
460	3365.7	2452.2	49.698	1240	11424.6	8962.1	59.668
480	3534.7	2581.5	50.058	1260	11661.0	9158.8	59.858
500	3706.2	2713.3	50.408	1280	11898.4	9356.5	60.044
520	3880.3	2847.7	50.750	1300	12136.9	9555.3	60.229
537	4027.5	2963.8	51.032	1320	12376.4	9755.0	60.412
540	4056.8	2984.4	51.082	1340	12617.0	9955.9	60.593
560	4235.8	3123.7	51.408	1360	12858.5	10157.7	60.772
580	4417.2	3265.4	51.726	1380	13101.0	10360.5	60.949
600	4600.9	3409.4	52.038	1400	13344.7	10564.5	61.124
620	4786.6	3555.6	52.343	1420	13589.1	10769.2	61.298
640	4974.9	3704.0	52.641	1440	13834.5	10974.8	61.469
660	5165.2	3854.6	52.934	1460	14080.8	11181.4	61.639
680	5357.6	4007.2	53.225	1480	14328.0	11388.9	61.800
700	5552.0	4161.9	53.503	1500	14576.0	11597.2	61.974
720	5748.4	4318.6	53.780	1520	14824.9	11806.4	62.138
740	5946.8	4477.3	54.051	1540	15074.7	12016.5	62.302
760	6147.0	4637.9	54.319	1560	15325.3	12227.3	62.464
780	6349.1	4800.1	54.582	1580	15576.7	12439.0	62.624
800	6552.9	4964.2	54.839	1600	15829.0	12651.6	62.783
820	6758.3	5129.9	55.093	1620	16081.9	12864.8	62.939
840	6965.7	5297.6	55.343	1640	16335.7	13078.9	63.095
860	7174.7	5466.9	55.589	1660	16590.2	13293.7	63.250
880	7385.3	5637.7	55.831	1680	16845.5	13509.2	63.403
900	7597.6	5810.3	56.070	1700	17101.4	13725.4	63.555
920	7811.4	5984.4	56.305	1720	17358.1	13942.4	63.704
940	8026.8	6160.1	56.536	1740	17615.5	14160.1	63.853
960	8243.8	6337.4	56.765	1760	17873.5	14378.4	64.001
980	8462.2	6516.1	56.990	1780	18132.2	14597.4	64.147
1000	8682.1	6696.2	57.212	1800	18391.5	14816.9	64.292
1020	8903.4	6877.8	57.432	1820	18651.5	15037.2	64.435
1040	9126.2	7060.9	57.647	1840	18912.2	15258.2	64.578
1060	9350.3	7245.3	57.861	1860	19173.4	15479.7	64.719

TABLE A-21E (*Continued*)

T	$\overline{h}$	$\overline{u}$	$\overline{s}°$	T	$\overline{h}$	$\overline{u}$	$\overline{s}°$
1900	19,698	15,925	64.999	3500	41,965	35,015	73.462
1940	20,224	16,372	65.272	3540	42,543	35,513	73.627
1980	20,753	16,821	65.543	3580	43,121	36,012	73.789
2020	21,284	17,273	65.809	3620	43,701	36,512	73.951
2060	21,818	17,727	66.069	3660	44,280	37,012	74.110
2100	22,353	18,182	66.327	3700	44,861	37,513	74.267
2140	22,890	18,640	66.581	3740	45,442	38,014	74.423
2180	23,429	19,101	66.830	3780	46,023	38,517	74.578
2220	23,970	19,561	67.076	3820	46,605	39,019	74.732
2260	24,512	20,024	67.319	3860	47,188	39,522	74.884
2300	25,056	20,489	67.557	3900	47,771	40,026	75.033
2340	25,602	20,955	67.792	3940	48,355	40,531	75.182
2380	26,150	21,423	68.025	3980	48,939	41,035	75.330
2420	26,699	21,893	68.253	4020	49,524	41,541	75.477
2460	27,249	22,364	68.479	4060	50,109	42,047	75.622
2500	27,801	22,837	68.702	4100	50,695	42,553	75.765
2540	28,355	23,310	68.921	4140	51,282	43,060	75.907
2580	28,910	23,786	69.138	4180	51,868	43,568	76.048
2620	29,465	24,262	69.352	4220	52,456	44,075	76.188
2660	30,023	24,740	69.563	4260	53,044	44,584	76.327
2700	30,581	25,220	69.771	4300	53,632	45,093	76.464
2740	31,141	25,701	69.977	4340	54,221	45,602	76.601
2780	31,702	26,181	70.181	4380	54,810	46,112	76.736
2820	32,264	26,664	70.382	4420	55,400	46,622	76.870
2860	32,827	27,148	70.580	4460	55,990	47,133	77.003
2900	33,392	27,633	70.776	4500	56,581	47,645	77.135
2940	33,957	28,118	70.970	4540	57,172	48,156	77.266
2980	34,523	28,605	71.160	4580	57,764	48,668	77.395
3020	35,090	29,093	71.350	4620	58,356	49,181	77.581
3060	35,659	29,582	71.537	4660	58,948	49,694	77.652
3100	36,228	30,072	71.722	4700	59,541	50,208	77.779
3140	36,798	30,562	71.904	4740	60,134	50,721	77.905
3180	37,369	31,054	72.085	4780	60,728	51,236	78.029
3220	37,941	31,546	72.264	4820	61,322	51,750	78.153
3260	38,513	32,039	72.441	4860	61,916	52,265	78.276
3300	39,087	32,533	72.616	4900	62,511	52,781	78.398
3340	39,661	33,028	72.788	5000	64,000	54,071	78.698
3380	40,236	33,524	72.960	5100	65,491	55,363	78.994
3420	40,812	34,020	73.129	5200	66,984	56,658	79.284
3460	41,388	34,517	73.297	5300	68,471	57,954	79.569

TABLE A-22E Ideal Gas Properties of Hydrogen, H_2

$T(°R)$, $\bar{h}$ and $\bar{u}$(Btu/lbmol), $\bar{s}°$(Btu/lbmol · °R)

$[\bar{h}_f° = 0$ Btu/lbmol$]$

T	$\bar{h}$	$\bar{u}$	$\bar{s}°$	T	$\bar{h}$	$\bar{u}$	$\bar{s}°$
300	2063.5	1467.7	27.337	1400	9673.8	6893.6	37.883
320	2189.4	1553.9	27.742	1500	10381.5	7402.7	38.372
340	2317.2	1642.0	28.130	1600	11092.5	7915.1	38.830
360	2446.8	1731.9	28.501	1700	11807.4	8431.4	39.264
380	2577.8	1823.2	28.856	1800	12526.8	8952.2	39.675
400	2710.2	1915.8	29.195	1900	13250.9	9477.8	40.067
420	2843.7	2009.6	29.520	2000	13980.1	10008.4	40.441
440	2978.1	2104.3	29.833	2100	14714.5	10544.2	40.799
460	3113.5	2200.0	30.133	2200	15454.4	11085.5	41.143
480	3249.4	2296.2	30.424	2300	16199.8	11632.3	41.475
500	3386.1	2393.2	30.703	2400	16950.6	12184.5	41.794
520	3523.2	2490.6	30.972	2500	17707.3	12742.6	42.104
537	3640.3	2573.9	31.194	2600	18469.7	13306.4	42.403
540	3660.9	2588.5	31.232	2700	19237.8	13876.0	42.692
560	3798.8	2686.7	31.482	2800	20011.8	14451.4	42.973
580	3937.1	2785.3	31.724	2900	20791.5	15032.5	43.247
600	4075.6	2884.1	31.959	3000	21576.9	15619.3	43.514
620	4214.3	2983.1	32.187	3100	22367.7	16211.5	43.773
640	4353.1	3082.1	32.407	3200	23164.1	16809.3	44.026
660	4492.1	3181.4	32.621	3300	23965.5	17412.1	44.273
680	4631.1	3280.7	32.829	3400	24771.9	18019.9	44.513
700	4770.2	3380.1	33.031	3500	25582.9	18632.4	44.748
720	4909.5	3479.6	33.226	3600	26398.5	19249.4	44.978
740	5048.8	3579.2	33.417	3700	27218.5	19870.8	45.203
760	5188.1	3678.8	33.603	3800	28042.8	20496.5	45.423
780	5327.6	3778.6	33.784	3900	28871.1	21126.2	45.638
800	5467.1	3878.4	33.961	4000	29703.5	21760.0	45.849
820	5606.7	3978.3	34.134	4100	30539.8	22397.7	46.056
840	5746.3	4078.2	34.302	4200	31379.8	23039.2	46.257
860	5885.9	4178.0	34.466	4300	32223.5	23684.3	46.456
880	6025.6	4278.0	34.627	4400	33070.9	24333.1	46.651
900	6165.3	4378.0	34.784	4500	33921.6	24985.2	46.842
920	6305.1	4478.1	34.938	4600	34775.7	25640.7	47.030
940	6444.9	4578.1	35.087	4700	35633.0	26299.4	47.215
960	6584.7	4678.3	35.235	4800	36493.4	26961.2	47.396
980	6724.6	4778.4	35.379	4900	35356.9	27626.1	47.574
1000	6864.5	4878.6	35.520	5000	38223.3	28294.0	47.749
1100	7564.6	5380.1	36.188	5100	39092.8	28964.9	47.921
1200	8265.8	5882.8	36.798	5200	39965.1	29638.6	48.090
1300	8968.7	6387.1	37.360	5300	40840.2	30315.1	48.257

TABLE A-23E Constants for the van der Waals, Redlich–Kwong, and Benedict–Webb–Rubin Equations of State

1. van der Waals and Redlich-Kwong: Constants for pressure in atm, specific volume in ft^3/lbmol, and temperature in °R

Substance	van der Waals		Redlich-Kwong	
	a $atm\left(\dfrac{ft^3}{lbmol}\right)^2$	b $\dfrac{ft^3}{lbmol}$	a $atm\left(\dfrac{ft^3}{lbmol}\right)^2(°R)^{1/2}$	b $\dfrac{ft^3}{lbmol}$
Air	345	0.586	5,409	0.4064
Butane (C_4H_{10})	3,509	1.862	98,349	1.2903
Carbon dioxide (CO_2)	926	0.686	21,972	0.4755
Carbon monoxide (CO)	372	0.632	5,832	0.4382
Methane (CH_4)	581	0.685	10,919	0.4751
Nitrogen (N_2)	346	0.618	5,280	0.4286
Oxygen (O_2)	349	0.509	5,896	0.3531
Propane (C_3H_8)	2,369	1.444	61,952	1.0006
Refrigerant 12	2,660	1.558	70,951	1.0796
Sulfur dioxide (SO_2)	1,738	0.910	49,032	0.6309
Water (H_2O)	1,400	0.488	48,418	0.3380

Source: Calculated from critical data.

2. Benedict-Webb-Rubin: Constants for pressure in atm, specific volume in ft^3/lbmol, and temperature in °R

Substance	a	A	b	B	c	C	α	γ
C_4H_{10}	7736.7	2587.6	10.26	1.9921	4.214×10^9	8.254×10^8	4.527	8.724
CO_2	562.3	702.4	1.850	0.7995	1.987×10^8	1.152×10^8	0.348	1.384
CO	150.6	344.1	0.675	0.8737	1.385×10^7	7.118×10^6	0.555	1.540
CH_4	203.0	476.0	0.867	0.6824	3.389×10^7	1.876×10^7	0.511	1.540
N_2	103.2	270.4	0.597	0.6526	9.700×10^6	6.700×10^6	0.523	1.360

Source: H. W. Cooper, and J. C. Goldfrank, *Hydrocarbon Processing,* 46 (12): 141 (1967).

TABLE A-24E Enthalpy of Formation, Gibbs Function of Formation, and Absolute Entropy of Various Substances at 537°R and 1 atm

$\overline{h}_f^\circ$ and $\overline{g}_f^\circ$ (Btu/lbmol), $\overline{s}^\circ$ (Btu/lbmol · °R)

Substance	Formula	$\overline{h}_f^\circ$	$\overline{g}_f^\circ$	$\overline{s}^\circ$
Carbon	$C(s)$	0	0	1.36
Hydrogen	$H_2(g)$	0	0	31.19
Nitrogen	$N_2(g)$	0	0	45.74
Oxygen	$O_2(g)$	0	0	48.98
Carbon monoxide	$CO(g)$	−47,540	−59,010	47.27
Carbon dioxide	$CO_2(g)$	−169,300	−169,680	51.03
Water	$H_2O(g)$	−104,040	−98,350	45.08
Water	$H_2O(l)$	−122,970	−102,040	16.71
Hydrogen peroxide	$H_2O_2(g)$	−58,640	−45,430	55.60
Ammonia	$NH_3(g)$	−19,750	−7,140	45.97
Oxygen	$O(g)$	107,210	99,710	38.47
Hydrogen	$H(g)$	93,780	87,460	27.39
Nitrogen	$N(g)$	203,340	195,970	36.51
Hydroxyl	$OH(g)$	16,790	14,750	43.92
Methane	$CH_4(g)$	−32,210	−21,860	44.49
Acetylene	$C_2H_2(g)$	97,540	87,990	48.00
Ethylene	$C_2H_4(g)$	22,490	29,306	52.54
Ethane	$C_2H_6(g)$	−36,420	−14,150	54.85
Propylene	$C_3H_6(g)$	8,790	26,980	63.80
Propane	$C_3H_8(g)$	−44,680	−10,105	64.51
Butane	$C_4H_{10}(g)$	−54,270	−6,760	74.11
Pentane	$C_5H_{12}(g)$	−62,960	−3,530	83.21
Octane	$C_8H_{18}(g)$	−89,680	7,110	111.55
Octane	$C_8H_{18}(l)$	−107,530	2,840	86.23
Benzene	$C_6H_6(g)$	35,680	55,780	64.34
Methyl alcohol	$CH_3OH(g)$	−86,540	−69,700	57.29
Methyl alcohol	$CH_3OH(l)$	−102,670	−71,570	30.30
Ethyl alcohol	$C_2H_5OH(g)$	−101,230	−72,520	67.54
Ethyl alcohol	$C_2H_5OH(l)$	−119,470	−75,240	38.40

Sources: Adapted from K. Wark, *Thermodynamics, 4th ed.,* McGraw-Hill, New York, 1983, as based on JANAF Thermochemical Tables, NSRDS-NBS-37, 1971; *Selected Values of Chemical Thermodynamic Properties,* NBS Tech. Note 270-3, 1968; and *API Research Project 44,* Carnegie Press, 1953.

INDEX TO FIGURES AND CHARTS

Figure A-1 Generalized compressibility chart, $p_R \leq 1.0$. *Source:* E. F. Obert, *Concepts of Thermodynamics,* McGraw-Hill, New York, 1960.

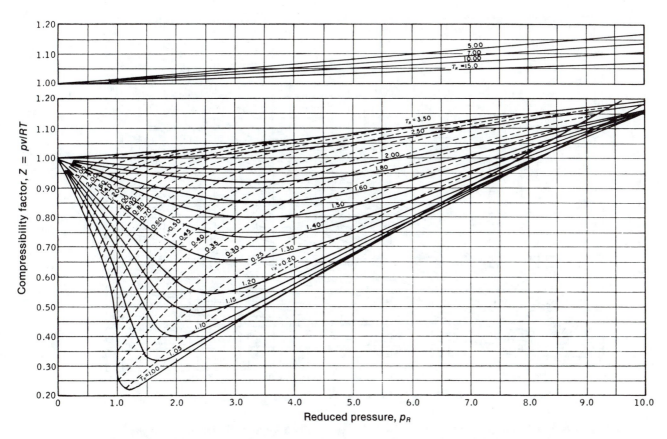

Figure A-2 Generalized compressibility chart, $p_R \le 10.0$. *Source:* E. F. Obert, *Concepts of Thermodynamics,* McGraw-Hill, New York, 1960.

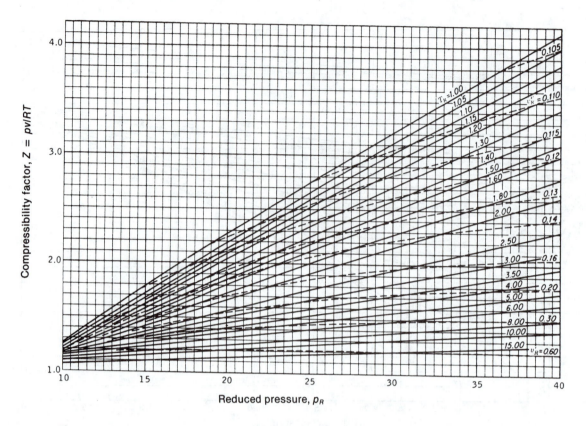

Figure A-3 Generalized compressibility chart, $10 \leq p_R \leq 40$. *Source:* E. F. Obert, *Concepts of Thermodynamics,* McGraw-Hill, New York, 1960.

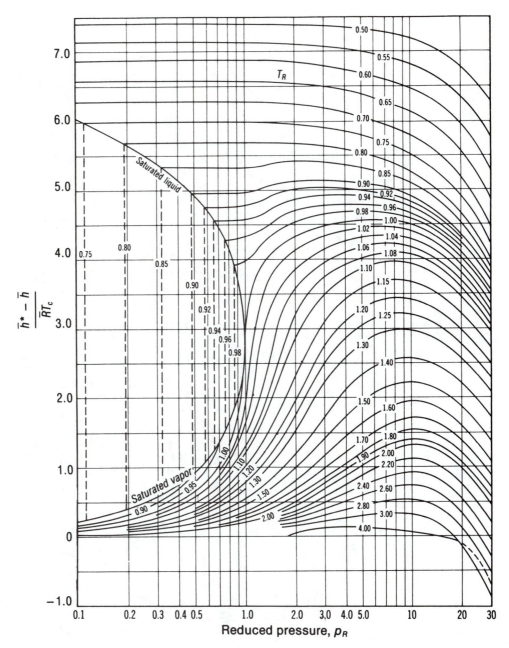

Figure A-4 Generalized enthalpy correction chart. *Source:* Adapted from G. J. Van Wylen and R. E. Sonntag, *Fundamentals of Classical Thermodynamics,* 3rd. ed., English/SI, Wiley, New York, 1986.

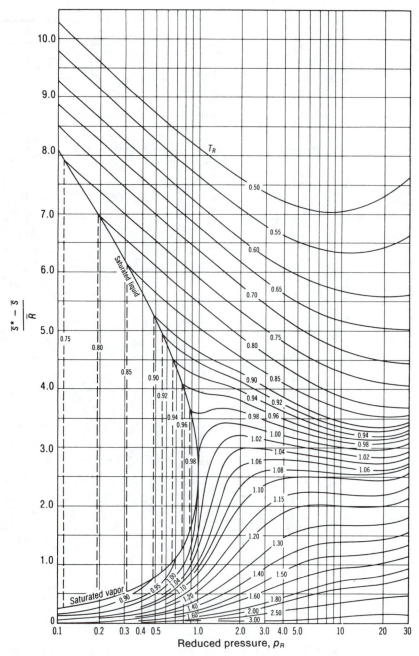

Figure A-5 Generalized entropy correction chart. *Source:* Adapted from G. J. Van Wylen and R. E. Sonntag, *Fundamentals of Classical Thermodynamics,* 3rd ed., English/SI, Wiley, New York, 1986.

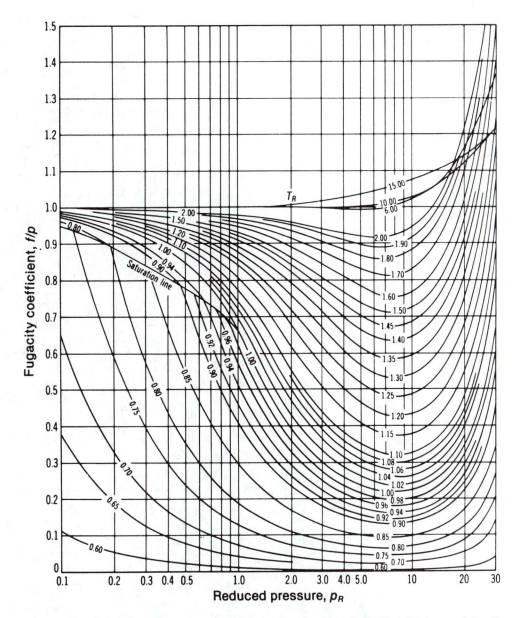

Figure A-6 Generalized fugacity coefficient chart. *Source:* G. J. Van Wylen and R. E. Sonntag, *Fundamentals of Classical Thermodynamics,* 3rd. ed., English/SI, Wiley, New York, 1986.

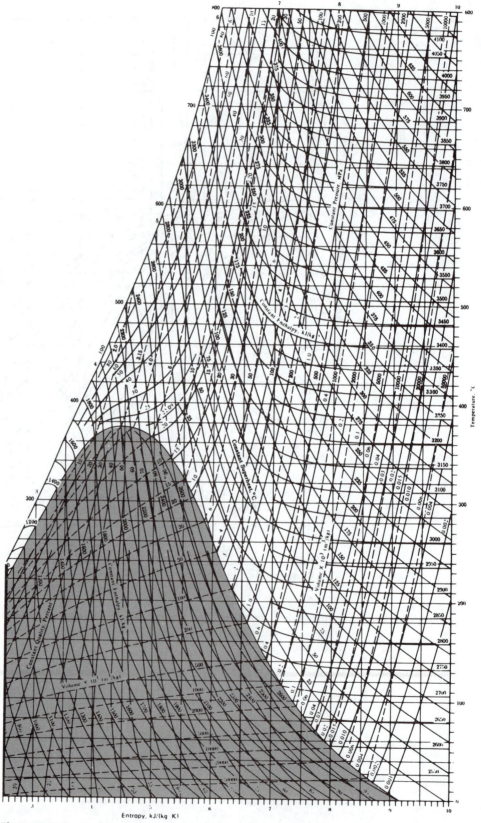

Figure A-7 Temperature–entropy diagram for water (SI units). *Source:* J. H. Keenan, F. G. Keyes, P. G. Hill, and J. G. Moore, *Steam Tables,* Wiley, New York, 1978.

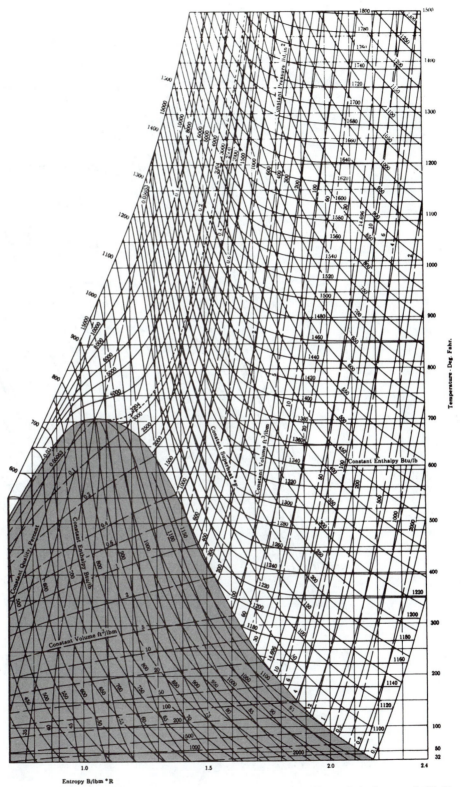

Figure A-7E Temperature–entropy diagram for water (English units). *Source:* J. H. Keenan, F. G. Keyes, P. G. Hill, and J. G. Moore, *Steam Tables,* Wiley, New York, 1969.

Figure A-8 Enthalpy–entropy diagram for water (SI units). *Source:* J. B. Jones and G. A. Hawkins, *Engineering Thermodynamics,* 2nd ed., Wiley, New York, 1986.

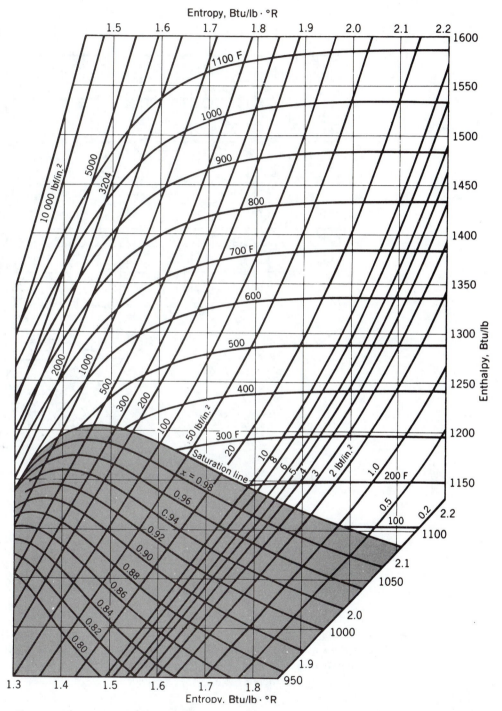

Figure A-8E Enthalpy–entropy diagram for water (English units). *Source:* J. B. Jones and G. A. Hawkins, *Engineering Thermodynamics,* 2nd ed., Wiley, New York, 1986.

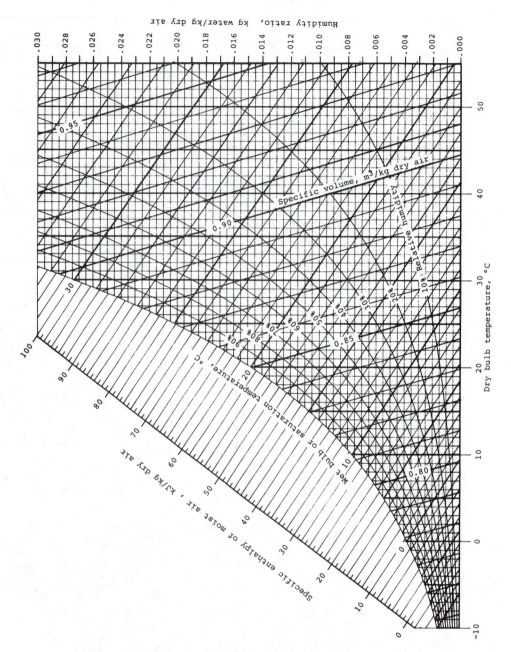

Figure A-9 Psychrometric chart for 1 atm (SI units). *Source:* Z. Zhang and M. B. Pate, "A Methodology for Implementing a Psychrometric Chart in a Computer Graphics System," *ASHRAE Transactions*, Vol. 94, Pt. 1, 1988.

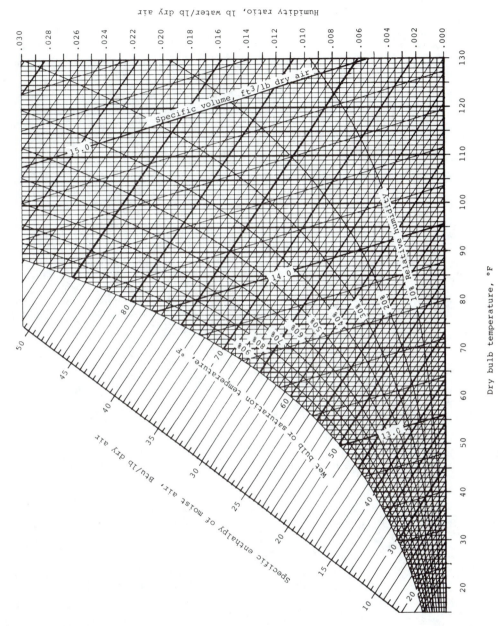

Figure A-9E Psychrometric chart for 1 atm (English units). *Source:* Z. Zhang and M. B. Pate, "A Methodology for Implementing a Psychrometric Chart in a Computer Graphics System," *ASHRAE Transactions,* Vol. 94, Pt. 1, 1988.